Tarzan was an Eco-Tourist ...
and Other Tales in the Anthropology of Adventure

Tarzan was an Eco-Tourist ... and Other Tales in the Anthropology of Adventure

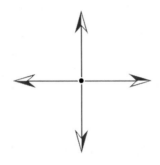

Edited by

Luis A. Vivanco and Robert J. Gordon

Berghahn Books
New York • Oxford

First published in 2006 by
Berghahn Books
www.berghahnbooks.com

©2006, 2009 **Luis A. Vivanco and Robert J. Gordon**
Paperback edition reprinted in 2009

Chapter 4, Universitetsforlaget (Scandinavian University Press) 1994;
Chapter 8 Sage Publications 2004

Library of Congress Cataloging-in-Publication Data

Tarzan was an eco-tourist— : and other tales in the anthropology of adventure /
edited by Luis A. Vivanco and Robert J. Gordon
    p. cm.
  Includes bibliographical references and index.
  ISBN 1-84545-110-4 — ISBN 1-84545-111-2 (pbk.)
    1. Adventure and adventurers. 2. Anthropology. I. Vivanco, Luis Antonio,
1969– II. Gordon, Robert J., 1947–

G525.T29 2006
910.4—dc22   **100669664X**
                                                    2006040260

British Library Cataloguing in Publication Data
A catalogue record for this book is available from the British Library

Printed on acid-free paper

ISBN 1-84545-110-4 hardback
ISBN 1-84545-111-2 paperback

Contents

List of Illustrations

Acknowledgements

This book has been an Adventure, in the Simmelian sense. That we deliberately set out on a trip of discovery is not what we, or Simmel, refer to here. For Simmel, the experience of adventure emerges in the unexpected and unplanned, the moments of dislocation where one can grasp the basic categories of existence. It is in this sense that putting together this volume has been adventuresome, because it has introduced us to new places and people that have encouraged us to rethink certain basic assumptions about the contemporary ubiquity of adventure, not to mention having altered some of the plans we laid out for this project.

The project was born of conversations, in the hallways between classes and in the hallways of conferences, about such things as television reality shows, the Ernest Shackleton craze on public television, the "extreme" content books and magazines one finds at the supermarket checkout counter, and our amazement that people actually buy such cars (and now cologne) as the HUMMER. All of it speaks to adventure's omnipresence, or at least a desire for adventure and its stories. As we expanded our conversations, inviting more friends and interlocutors, we found a lot of our colleagues had been thinking about these things as well, perhaps quite naturally, because as anthropologists some of our professional *mana* comes from the appearance that we lead adventuresome lives.

The time seemed ripe to convene more formal conversations about the complex relationships between culture and adventure, and so we organized two panel sessions, one at the American Anthropological Association annual meetings and another at the Northeastern Anthropological Association annual meetings. With the help of Rob Welsch, we also arranged a mini-conference in the Maori Room at the Field Museum in Chicago. Judging by attendance numbers and the animated conversations that continued as the rooms emptied, the topic struck a chord with a lot of people. We have seen a mixture of enthusiasm and disquietude, a sense that we have some things to work out as individuals and as a discipline about what this ubiquity of adventure tells us about the worlds in which we live and work.

Unfortunately, for reasons that can perhaps be explained only by Simmel's emphasis on the unexpected nature of adventure, a number of the friends who joined in these conversations could not join us in this volume. But we would like to express our deep gratitude to them, not only because of what they added to these conversations, but also because of the contributions they have made or are likely to make on this theme once they are able to publish the fascinating insights they shared with us. They include Ed Bruner, Quetzil Castañeda, Frederick Errington, Deborah Gewertz, Harald Prins, Mark Mosko, M. Estellie Smith, and Geoff White. Ute Luig, Lynn Meisch and John Middleton were also intellectually close to this project, although they did not in the end contribute to this volume. We also owe a gracious debt to the folks at Berghahn Books. Marion Berghahn has been supportive of the project from the very first moment we broached the topic with her. Catherine Kirby and Michael Dempsey have assisted expertly in the production phase.

At the University of Vermont, where this book finally came together, we have benefited from rich conversations with Guha Shankar and Glenn McRae while our Middlebury neighbor David Napier supplied enthusiasm and advice (but still owes us a few beers). We are also grateful to Ross Thomson, of our Economics Department, for providing us with background on the insurance industry, and Katie Weiss, a work-study student who impressed both of us with her editing skills.

We are especially grateful to our families, who consistently keep us in line but allow us to veer off on adventures of our own, often enough to keep us productively decentered.

Notes on Contributors

Alan Barnard is Professor of the Anthropology of Southern Africa at the University of Edinburgh. He has done fieldwork with the Nharo (Naro) of Botswana and research on the history of anthropological ideas. His books include *Hunters and Herders of Southern Africa* and *History and Theory in Anthropology.*

Daniel Bradburd is Professor and Chair of Humanities and Social Sciences at Clarkson University. He is co-editor of the book *Drugs, Labor and Colonial Expansion.* He has worked in Southern Iran with Komachi nomads.

Rob Gordon is affiliated with the Free State University and Professor of Anthropology at the University of Vermont. He has done fieldwork in southern Africa and Papua New Guinea. His books include *Law and Order in the New Guinea Highlands; The Bushman Myth and the Making of a Namibian Underclass;* and *Picturing Bushmen: The Denver African Expedition of 1925* as well as several edited volumes.

David Houston is a Doctoral Candidate at McGill University. He teaches in the Anthropology department at the University of Vermont, where he also works in computing.

Laura Hubbard is a PhD candidate in cultural anthropology at the University of California, Berkeley. She is currently writing a dissertation that examines the traffic in media forms between Johannesburg and Lusaka, emphasizing how regional relations of power unsettle commonly held notions of cultural globalization in Southern Africa.

Lamont Lindstrom is a professor of anthropology at the University of Tulsa. He is the author of *Cargo Cult: Strange Stories of Desire from Melanesia and Beyond* (University of Hawaii Press, 1993) and *Knowledge and Power in a South Pacific Society* (Smithsonian, 1990); coauthor of *Kava: The Pacific Drug* (Yale University Press, 1992) and *Island Encounters: Black and White Memories of the Pacific War* (Smithsonian, 1990); and has also published on kava, chiefs and governance, and language on Tanna (Vanuatu).

Joy Logan is a member of the Spanish Division in the Department of Literatures and Languages of Europe and the Americas at the University of Hawai'i where she teaches Latin American and Latino literature and cultural studies. She has written on redemocratization, postmodernism, feminism, life-writing, and women's writing workshops in Argentina. Her current research on the development of mountaineering on Aconcagua focuses on globalization and the local and representations of gender, indigeneity, and regional identity in the expansion of mountain tourism.

Kathryn Mathers was a postdoctoral fellow in the Department of Anthropology and Archaeology at the University of Pretoria, South Africa, and is currently a Teaching Fellow at Stanford University. She is interested in the ways that American-ness is constructed through interactions between the U.S. and Africa and is writing about the impact on Americans of travel to southern Africa.

Keally McBride is Senior Fellow, University of Pennsylvania, Department of Political Science and the Center for Programs in Contemporary Writing. She is interested in how illusion and fiction shape political institutions and choices. She has two books forthcoming, *Collective Dreams: Political Imagination and Community* (2005) and *Punishment and Political Order* (2006).

A. David Napier teaches anthropology at University College London, and is the Director of Students of Human Ecology, a registered nonprofit dedicated to applied learning in the areas of health care and indigenous rights. He is the author of four books on the cultural construction of self and oversees a number of applied projects, including Public Anthropology's (www .publicanthropology.org) "Network for Student Activism."

Jason J. Price is a McCracken Fellow in the Department of Anthropology at New York University.

Steven Rubenstein is a member of the Institute of Latin American Studies at the University of Liverpool. He is the author of *Alejandro Tsakimp: A Shuar Healer in the Margins of History*. He is currently developing projects on eco-tourism among the Shuar, and Shuar involvement in national politics in Ecuador.

Michael Sheridan teaches cultural anthropology at Middlebury College. He was a Peace Corps Volunteer in Kenya in 1988 to 1990. His research focuses on the intersection of material and symbolic processes in African resource management.

David Stoll teaches anthropology at Middlebury College. He is the author of *Is Latin America Turning Protestant?*; *Between Two Armies in the Ixil Towns of Guatemala*, and *Rigoberta Menchú and the Story of All Poor Guatemalans*.

Luis Vivanco is Associate Professor of Anthropology at the University of Vermont. His research is on the cultural politics of environmentalism, eco-tourism, and sustainable development in Costa Rica and Mexico. He is author of *Green Encounters: Shaping and Contesting Environmentalism in Rural Costa Rica* (Berghahn Books, 2006).

Neil L. Whitehead is Professor of Anthropology at the University of Wisconsin-Madison. He has authored numerous works on the travel literature, history, and anthropology of South America; his most recent volumes include *In Darkness and Secrecy. The Anthropology of Assault Sorcery and Witchcraft in Amazonia* (Duke University Press, 2004), *Nineteenth Century Travels, Explorations and Empires, 1835–1910; South America* (Chatto & Pickering, 2004), and *Histories and Historicities in Amazonia* (University of Nebraska Press, 2003).

Aram A. Yengoyan (PhD, University of Chicago, 1964) taught in the Department of Anthropology, University of Michigan from 1963 to 1990 at which time he joined the Department of Anthropology, University of California, Davis. He has done his primary fieldwork among upland peoples of southeast Mindanao, Philippines, and also among the Pitjantjatjara of the western desert of Australia. His theoretical interests and writings deal with cultural theory, ideology, language and culture, and translation.

CHAPTER 1

Introduction

Robert J. Gordon

We live in a post-explorer era in which it is widely considered that the feats of the great adventurers are remnants of history and that the Earth's mysterious places and peoples have long "been discovered." Yet adventure enjoys ubiquitous status in public culture and late capitalism. Adventure television, from the Discovery Channel to the "reality shows," is a major growth area. Best-selling books and magazines increasingly feature "extreme content" and narratives of audaciously successful and famously disastrous expeditions. The best selling SUV (Sport Utility Vehicle) speaks volumes about the current fascination with adventure and the goods deemed necessary for it. Such purchases clearly have major environmental consequences, yet people persist in purchasing them even though they—like most SUVs—will never be used for what they are supposed to be capable of doing. These totems of the desirable bespeak a nostalgia for more heroic days.

Or consider "adventure travel." Once the province of elites, it has become accessible and fashionable among the middle classes, and is one of the fastest-growing segments of the tourism industry. It draws on, and serves as a conduit for, an increasingly transnational concern with the disappearance of distinctive places, cultures, and ecosystems (Johnston 1990; Zurick 1995). More than half of tourists (98 out of 197.7 million) defined themselves as adventure travelers (Travel Industry Association 1998).

Adventure should not of course be restricted to consumers: its therapeutic value is well-accepted. Over the last three decades, for-profit and not-for-profit agencies have developed adventure experiences defined as life-enhancing and identity-transforming. The best known of these organizations is Outward Bound, which has established a significant market niche by emphasizing the power of personal challenge. Outward Bound, like most experiential programs such as Ropes and NOLS (National Outdoor Leadership School), embraces the belief that personal experience heightens self-awareness and builds "character" (Holyfield & Fine 1997).

Although anthropologists have explored kindred topics like danger and risk in depth (e.g. Douglas and Wildavksy 1982; Caplan 2000; Lianos with Douglas 2000), few have explicitly considered adventure as a subject in its own right, much less tried to explain the contours of its ubiquity for contemporary cultural production. There are good reasons for anthropology to examine contemporary concepts and forms of adventure, and this volume is an initial effort at presenting some of the major themes that could fall under an anthropology of adventure. We ask: What does it mean to have an "adventure," experientially speaking? What conditions transform quotidian lives and activities into adventure? How do differential access to resources or one's position in social hierarchies of gender, race, ethnicity, sexuality, etc. affect perspectives on risk, danger, and the experience of adventure? What is the social organization of adventure? What are the different ways adventure is commodified and visualized? How do Western understandings of adventure translate cross-culturally? What rituals and symbols are associated with contemporary forms of adventure? How does anthropology itself reflect shifting concepts and practices of adventure?

In this introduction, we examine the meanings of adventure, using as a focal point the seminal work of Georg Simmel. We then briefly situate the historical appearance and rise of a certain kind of adventurer and adventuresome sensibility that is captured in the imagery of Tarzan. We also explore the social conditions that have helped shape modern adventuring, particularly colonial-era traveling, the invention of the camera, and the rise of cinema. We examine contemporary expressions of adventure, and explore their relationship to commodification and capitalism. We conclude with an exploration of anthropology's own ambivalence toward adventuring, which derives from a combination of fascination and professional distancing.

Defining Adventure

The Oxford Concise Dictionary defines adventure as: "Risk, danger; daring enterprise; unexpected incident," and then adds "commercial speculation."

As if to underline this latter point, it defines an "Adventurer" not only as "one who seeks adventures," but also a "soldier of fortune, speculator, one who lives by his wits." A crucial element of adventure, as Nick Thomas has observed, is the "reconstitution of the landscape, certain things and some social relations in new imaginative terms" (Thomas 1987: 10). As a form of self-expression its potential lies in its ability to shift "between real and metaphorical contexts of deterritorialization" (Wardle 2002: 525). Adventures are fueled in the imagination but grounded in perceived and real risk that, were it not for the adventure, could be avoided. It is a special moment and place, as Erving Goffman (1967) puts it, "where the Action is." One has to be prepared, momentarily at least, to let go of a controlled situation and accept one's fate.

In Western cultural contexts, adventure is often communicated and understood through a biochemical idiom, as the pursuit of the "adrenaline rush" and the "endorphin high." Recent research does indeed show that risk-taking can have a drug-like effect on people because it releases dopamine, the chemical transmitter that pushes the neurological levers marked "gratification" in the mesolimbic reward system (Dabbs 2000). Charles Pasternak (2003) argues that what differentiates humanity from our genetically close cousins is the synergy of three otherwise ordinary evolutionary vectors, namely the development of the opposable thumb, a slight alteration in vocal-cord anatomy which allowed for a broad range of sounds, and the rapid accretion of neurons in the brain's cortex, all of which fueled a "taste" for searching, analyzing, and exploring. In short, questing. Indeed, some psychologists see evolutionary advantage in gratuitous risk-taking, in that it promotes both physical and cultural exploration and growth (see e.g., Apter 1992; Zuckerman 1994).

Adventure arguably has a close relationship to human biology, but to reduce adventure to its purely biochemical and evolutionary manifestations misses the rich social, cultural, political, and economic contexts that shape how and why people think of certain activities and images as adventuresome. Clearly a focus on adventure can be an important way in which to approach some of our central interests as anthropologists. Adventure provides a useful context in which to study issues of political economy, development, colonialism, and globalization, as well as issues of cultural identity, subjectivity, and representation, precisely because it breaks from, and can be a useful vantage point from which to consider, the humdrum of the ordinary life.

Simmel is one of the few social theorists to analyze adventure and his work provides an important thread linking all the chapters in this volume. In his classic 1911 essay "The Adventure," he observes that adventures are experiences that occur beyond the humdrum of everyday life. Adventures have clearly demarcated beginnings and endings, and generally entail dropping out of the continuity or reciprocal interpenetration with adjacent parts of what constitutes "life as a whole." As an experience, it is like "an island of life which determines its beginnings and end according to its own formative powers and—like the part of a continent—also according to those adjacent territories" (Simmel 1983: 223). Being free of normal entanglements, an adventure has a quality of self-sufficiency. They are part of the *exclave* of life, "torn-off" from a (somehow) unified stream of existence. Adventures thus frequently take on a dreamlike quality, which provides a heightened, if ahistorical, form of experiencing (Simmel 1983: 222–226). Both Yengoyan's and Gordon's chapters in this volume note the antistructural characteristics and similarities to some of Victor Turner's work in this regard. Adventures typically occur in non-normal places, that is, adventurers are by and large "strangers," another category famously theorized by Simmel (1950). In such untrustworthy places stranger/adventurers have a strong inducement to capture what impresses them most as "authentic" and "typical."

Risk-taking is intrinsic to adventure. When asked why he climbed mountains, Sir Chris Bonington replied:

> It's the drive to adventure.... To give it a wider connotation, if human kind did not get challenged by risk, we'd probably still be in the caves or probably would never have got where we were. So I think an intrinsic important part of the human psyche is this desire to stretch themselves to go into the unknown (BBC interview with Robin Lustig, Broadcast 29 May, 2003)

For Bonington mountaineering is a "calculated risk." The thrill lies in challenging danger and then using skill to obviate it. The word "risk" derives from the Italian *risicare,* "to dare." It is thus not a fate, but a choice that will depend on how much knowledge one has and how free one is to exercise options. Risk is therefore a cultural construction that is the result of active agency, not some passive reaction. As Paine (2002: 68) puts it, danger is objective, and risk is how one interprets it. Risk perception is thus shaped by both culture and social organization. Within such a social cosmos the distribution of power will obviously influence such perceptions (Douglas and Wildavsky 1982). As Simmel noted, adventure entails "the gesture of the conqueror." Inevitably adventurers have considerable coercive power, in that in pursuing their objectives they do not have to consider the objective natural tendencies. They labor under the apprehension that they can successfully battle the elements and conquer nature. Adventures typically occur in risky environments that would prod adventurers to be more careful, and indeed, calculated in their actions. In a study of mountaineers, the common characteristic to emerge was a strong attraction to ambiguity, uncertainty, and puzzlement (Mitchell 1998). They perceived themselves not as boisterous devil-may-care adventurers, but rather as capable people who enjoyed meeting challenges with complex problem-solving skills. Mountains are the classic site for adventuring, and several chapters in this volume, those by Houston, Logan, and Napier, deal with facets of this "sport."

Risks are generally based on imponderables that are frequently a product of lack of knowledge or the need to uncover a secret, a topic also pioneered by Simmel (1950). Awareness of secrets produces an "immense enlargement of life." An adventure thus involves undertaking action(s) involving dangers and risks the extent of which is "unknown" or a still a secret, which ultimately possibly imperil one's own existence. This element of awareness of danger is crucial. In his reminiscences, Sartre recalled that it was while serving in the French Underground against the Nazis that he felt most free: "Freedom is Terror and Terror is Freedom," he famously proclaimed. His countryman Camus said it well: "What gives value to travel is fear. It breaks down a kind of internal structure … stripped of all our crutches, deprived of our masks … we are completely on the surface of ourselves … this is the most obvious benefit of travel" (Camus 1962: 26). This is why a sense of adventure is so necessary for imperialism since the place of adventure is usually beyond what is defined as home or "civilization." Here nothing can be trusted, nothing can be taken for granted.

To be sure, people take risks all the time, but that does not necessarily mean that they are having an adventure, at least if we follow Simmel. One of the characteristics of modernity, as Goffman (1968) points out, is that chance-taking tends to be an organized affair, as in a competitive commercialized sport or gambling. Casinos are certainly places where people take risks; but because they are specially organized social spaces, adventures rarely happen within them. There are, of course, also a number of jobs or careers that hold the potential for adventure. Indeed, a popular euphemism for mercenaries is "professional adventurers," and one simply has to look, as McBride does in this volume, at current U.S. military recruiting strategies to see how jobs are "adventurized." But even these high-octane jobs can be incredibly boring most of the time (Goffman 1968; Bourke 1999). In Simmel's formulation these would not be adventures as they do not entail time away from ordinary work. Adventure entails a strong ludic and avowedly nonwork element because it deals with nonessential chances. Indeed, for Michael Green, adventure is freedom to engage in normally socially unacceptable behavior. Being outside the normal bonds of society, one can give satisfaction to the wilder impulses and desires (Green 1993: 185). Gordon's chapter in this volume addresses this issue by comparing South African troops and United Nations peacekeepers in the same environment and argues that the success of the latter can be attributed to the fact that they saw their deployment as an adventure within a structured *communitas* situation, while the former did not.

For Simmel, adventure is an experiential framework: "The decisive point is that the adventure, in its specific nature and charms, is a form of experiencing. The content of the experience does not make the adventure" (1983: 229). Michael Crichton, the science fiction writer, describes how in an adventure one is "stripped of your ordinary surroundings, your friends, your daily routines … you are forced into direct experience" (cited in Strain 2003: 4). Adventures, it would seem, peel away one's own cultural baggage. At the root of tourism, and especially adventure, some suggest, is the quest for an authentic experience. Showing shades of Simmel's (1950) work on the "metropolis," MacCannell (1989) suggests that this is a reaction to urban alienation that grew out of industrialization and fragmentation of modern society. Authenticity, or the "cult of immediate experience" (Conner 1978: 56), is becoming increasingly important in this globalized world. Modern tourists and adventurers fervently pursue authenticity that they believe can be found by stripping away layers of culture and is a quality that so-called "primitive" societies experience. By subjecting their bodies to risk and stress in untamed nature and cultures, many Westerners believe that not only are they "getting away from it all" but they are also getting "in touch with their real selves."

This raises an important problem, Bruner (1986) argues, namely, that we can only experience our own lives, although we may seek clues and make inferences about another's experiences, typically by interpreting expressions.

To be a successful adventurer one must be a storyteller as well, and such stories are complex, their success depending on their ability to "transport" the audience so that they can live vicariously through the tale. The true mark of a hero or adventurer, Hughes-Hallett (as cited in Holland 2004) has recently suggested, is the ability to inspire in others forms of madness— whether desire, or terror, or both. Of course part of the art of adventure, as Bradburd and Napier point out in this volume, is precisely in denying the heroic. It requires more than simply rehashing the act and needs to include the "subtleties of feeling which flesh out 'mere' acts" (Zweig 1981: 96). Stories of danger can stimulate vicarious pleasures. Thus when a German tourist was mauled by a lion in the Etosha Game Park, tourism authorities expected it to have a negative impact on numbers but the story had precisely the opposite effect, leading to a healthy increase in tourism.

It is also necessary to distinguish adventure from other types of related experience, namely tourism and pilgrimages (although Houston's essay in this volume suggests that in mountaineering they can merge). Central to the marketing of the travel and leisure industry is the notion of an escape from the everyday. Most religions encourage travel in the belief that it is good for the soul. Pilgrimages focus on self-transformation. They are experienced and interpreted as rebirth, atonement, or liberation from materialism, jealousy, and hatred (and also of course status enhancement). While historically pilgrims were undoubtedly transformed mentally and physically, nowadays "time-space compression" has resulted in pilgrimages and adventures becoming vulgarized. One has to interact, albeit transiently, with strangers, thus the standard norms of reciprocity do not apply and in this sense it can be a high-risk activity. But what organized mass tourism does is minimize these risks by taking "the trouble out of traveling." The power and economic differential is such that tourists can typically withdraw if they feel threatened. And then there is that hybrid, adventure tourism. It entails a contradiction since adventure is about dealing with uncertainty, yet planned tours minimize this. Adventure tourism is marketed for those who have neither the time nor the desire to take the risk fully upon themselves.

Reminiscing about their adventures, it is clear as Houston's chapter suggests, that what many adventurers most fondly remember is the intense camaraderie, in which they rely upon one another not only for companionship, but for support in life-and-death situations. Indeed in the prefeminist era, Lionel Tiger wrote an influential book *Men in Groups* (1969), in which he argued that adventures were necessary for male-bonding, and that male-bonding was critical for maintaining social order. In the Western imagination, adventurers have been almost exclusively male, and the very term carried with it a certain air of gallantry, while the term "Adventuress" brought forth mostly negative connotations around the theme of "gold digger." But this is changing. Nowadays adventure provides the grounds upon which conflicts over gender equality play out, as Logan and Mathers and Hubbard's chapters suggest.

Given the experiential component of adventure, it is not surprising that some (including Simmel himself) claim that sex is the signature adventure, since each sexual encounter is believed to be unique (Cohen and Taylor 1992). It is no accident that the only adventurer Simmel mentions by name is Casanova! Part of this dynamic undoubtedly is that the imagination often links cultural and sexual motives for travel. Sex for liberation and revelation was often an unstated feature of European travel in the nineteenth and twentieth centuries. As Sir Richard Burton, an early adventurer-anthropologist, recognized, a change of place frequently signaled a change of morals. There is a common belief that the "other" is always promiscuous. Thus Americans and Europeans believe that Africans and Latin Americans are licentious (Gilman 1985), while Africans, Arabs, Asians, and Latin Americans commonly believe the same of Europeans and Americans (Sumich 2002). It is not only men who engage in sexual tourism but increasingly also women, as Mathers and Hubbard suggest (see also Ebron 1997; Meisch 1996, 2002). Whether or not travelers actually engage in sex while abroad, sexual encounters often feature prominently in tales and fantasies, not only of the adventurers, but also of the local people on the site of the adventure (Bowman 1989; Sumich 2002).

Sex is frequently a coy part, at least of the literature on the adventure of travel, yet the sensual is an important if overlooked part of the adventure. In his essay on "Flirting," Simmel (1984) captures this corporeal ambience well. Flirting, he says, is a way of being suspended between having and not having, between consent and refusal. Flirting captures much of the contemporary sense of adventure with its air of self-indulgent lack of responsibility. But like yearning, it ceases once it is fulfilled. One can only be said to have flirted in the afterglow, once the danger has receded. The key quality is that of sustained ambiguity. It is this desire for sustained ambiguity implicit in adventure that recently led Laura Kipnis to pen a polemic entitled *Against Love*. She argued in favor of adultery because it allowed one to step outside the bounds of what people normally do. It discombobulates temporality and provides a feeling of elation coupled to anxiety and gnawing guilt. As she says, "A strange virus seems to have invaded your normally high-functioning immune system, penetrating your defenses, leaving you vulnerable, trembly, strangely flushed. It seems you have contracted a life-threatening case of desire" (Kipnis 2003: 7). Adultery captures the metaphysics of adventure well. One is pushing the limits. One is betwixt and between statuses, and it could lead to a change in status. No wonder it holds the potential to be transforming, yet by definition militates against transformation.

Tarzan Ascendant

Sex is a powerful frame for considering adventure. But in this volume, we self-consciously turn to a more macho image, that of Tarzan. We do this because we feel he provides an even more powerful lens to consider adventure's ubiquity. As an icon Tarzan both transcends and highlights the specificity of adventure's historical meanings. In their day, Edgar Rice Burroughs's novels enjoyed enormous popularity because Tarzan represented the consummate colonial-era adventurer: a white man whose noble civility enabled him to communicate with and control savage peoples and animals. Tarzan is now the consummate "eco-tourist": a cosmopolitan striving to live in harmony with nature, using appropriate technology, and helping the natives who are too dumb to solve their own problems.[1] Tarzan is still an icon of adventure, because like all adventurers, his actions have universal qualities (observations that both Napier and Vivanco explore). But the meanings assigned to his adventurous actions, as with any adventure, are also highly dependent on specific historical, cultural, and political contexts. So what does Tarzan stand for and how do we explain his longevity and appeal?

Burroughs, who had never been to Africa, wrote his Tarzan books with a "50 cent Sears dictionary and Stanley's *In Darkest Africa*" (Torgovnick 1990: 26). He knew little about Africa and cared even less. His novels, and the movies based on them, are a hodgepodge of generic fantasies about savages and jungles, frequently not even based in Africa, and no one, certainly not the reviewers, bothered too much about such matters (Fury 1994). Tarzan clones like *Jacaré* also worked on this formula, as Whitehead shows. At its peak Burroughs's negative portrayal of women and blacks did not appear to attract criticism. Yet, according to George MacDonald Fraser, "probably no colonial writer except Haggard so shaped Western imagination of Africa and its people" (1988: 142).

Burroughs's first Tarzan novel, *Tarzan of the Apes,* was published in 1912 and a stream of twenty-three others followed. So successful was this enterprise that Burroughs registered Tarzan as a trademark in 1913 and licensed everything from coffee and bread to bubble gum and toys. Tarzan was the first fictional character to be multi-mass-media marketed. A Tarzan newspaper comic strip was successfully launched in 1929 (Green 1992: 199), and radio shows followed in 1931 (Morton 1993: 106). There was a veritable Tarzan industry, with "Tarzan Clubs" rivaling the Boy Scouts for the loyalties of boys and young adult males. But Tarzan influenced more than young boys. He was popular globally, translated into more than thirty-two languages. As late as 1963, Tarzan constituted one-thirtieth of the total annual sales of all paperbacks in the U.S. (Mandel 1963; Torgovnick 1990: 42). Of course the heady sixties also heralded the age of the Boeing, the rise of global mass air travel, and more importantly the founding of the Peace Corps and similar "helping agencies." Now instead of simply reading and fantasizing about Tarzan, one could emulate him, a theme that Sheridan and Price examine in their analysis of the Peace Corps as adventure.

Commentators like Gore Vidal (1963) interpreted Tarzan's viselike hold on popular memory to a desire and fantasy to escape, to dominate the environment. But such a reading, as Torgovnick (1990) suggests, is elitist, presenting the folks of the 1920s as dupes. Rather, she argues, Tarzan promised not an escape but a recreation of the modern world. The messages emanating from his novels were complex. He was a mover and a shaker who made things happen, like in the Peace Corps hype, or (as Vivanco suggests) the television figure of the Crocodile Hunter.

By the turn of the century the American frontier had been pronounced closed and the appreciation of wilderness reached national cult proportions, spearheaded by such popularizers as Teddy Roosevelt and John Muir. Many believed that white masculinity was under threat from a variety of factors, including the fledgling suffragette movement. Masculinity could be retained and regained by physical work and contact with nature. Tarzan's enduring popularity, Kasson believes (and Vivanco echoes), testifies to the enduring importance of manliness confronting the wilderness: Life is too soft and one has to prove and test oneself physically in tight spots. At the same time Tarzan represents a crude, albeit effective, intervention in the nature-nurture debate on the popular front, coming out clearly for "good breeding" (Kasson 2001; Bloom 1993), or the "Blue bloods" as Napier would have it. The interrelations between Tarzan, adventure, and the origins of anthropology and "popular" culture were complex, and cry for detailed analysis, and the chapters by Yengoyan, Barnard, Bradburd, and Lindstrom examine various aspects of this situation.

Tarzan's Lineage: The Rise of Exhibitionary Adventurism and Modern Adventure Tourism

> Nowadays, being an explorer is a trade, which consists not ... in discovering hitherto unknown facts ... but in covering a great many miles and assembling lantern-slides or motion pictures ... so as to fill a hall with an audience for several days in succession (Levi-Strauss 1974: 4).

The rise of modern adventuring is located in the aftermath of the Napoleonic Wars. Not only did this herald a period of comparative peace in Europe, it also signaled the growing importance of industrialization with all its important cultural consequences. There was also a powerful, if neglected, demographic stimulus. Demobilization was complex. The ordinary soldiers and sailors were easily discharged, but officers were a more politically volatile matter since they had the vote and political connections. Instead, they went on "half-pay," leaving the British Royal Navy, for example, with a ratio of one officer for every four sailors. In 1846, of 1,151 officers, only 172 were in full employment (Fleming 1998: 2). In such a situation "exploration" held a special appeal to the middle classes, and recently established organizations like the Royal Geographical Society helped to stimulate, channel, and advise on such matters. Gordon Cummings, an ex-Indian Army officer who pioneered sports adventure hunting in Africa, epitomized their approach and attitude. Cummings outsold Charles Dickens with his *The Lion Hunter of South Africa,* and claimed to be richer than Britain's wealthiest landowner, for in Africa "I felt that it was all my own" (Bull 1988: 64).

And then there was David Livingstone, still one of the European names most closely connected to Victorian-era adventure. How he achieved this near-iconic status informs us about the persistence of certain imageries of the Other and the sense of adventure they evoked. His first book, *Missionary Travels and Researches in South Africa,* was a best seller and sold over 70,000 copies. Changes in lithography and the emergent "penny presses" ensured that illustrated books like those by Livingstone enjoyed a wide and rapid distribution, not just among the upper classes but, more importantly, among the aspiring classes as well. Given the inherently proselytizing nature of his mission work both in Africa and at home (perhaps one of the more successful examples of the use of "adventure" for mass fund-raising) he soon became a household name, especially when he got "lost." In 1871, in one of the greatest and most celebrated scoops in journalism, *The New York Herald*'s man, H.M. Stanley, met and interviewed him. Stanley's account, *In Darkest Africa* (150,000 copies), was read, in the words of one 1890 reviewer, "more universally and with greater interest than any other publication" (cited in Brantlinger 1988: 180). This Boorstinian "pseudo-event" was one of the milestones in the history of the mass media. Changes in print technology unleashed heavy competition, especially in the United States and Britain among what was later dubbed the "Yellow Press," and served to create and sustain the myth of the Explorer, epitomized by the remarkable success of the slew of self-promoting books Stanley later wrote (Riffenbaugh 1992).

Directly inspired by the tales of Livingstone was Francis Galton, cousin to Charles Darwin, founder of eugenics and second President of the Royal Anthropological Institute. What is less well-known about Galton is that as a twenty-five-year-old, emboldened by Livingstone's "discovery" of Lake Ngami, he tried to be the first European to reach the lake from the West. While he failed in this endeavor, within a year of his return to England he published a book, *Tropical South Africa,* for which he was awarded the Gold Medal of the Royal Geographical Society, and the Silver Medal of the French Geographical Society, and was elected to both the Royal Society and the Athenaeum Club. The book went through at least four editions and was translated into German and French. Buoyed by the success of this effort, he immediately started work on a second book that was also based on his travels. This was his even more successful best seller *The Art of Travel, or, Shifts and Contrivances Available in Wild Countries,* first published in 1855 and eventually running to eight editions. Clearly, its sales were to be found not so much among the emergent professional explorers who found it rather fanciful, as among the "armchair" travelers, and it informed and sustained the Victorian visual imagery of Africa and other exotic lands. It literally and figuratively set the standard for what adventures were supposed to be. This is clear from the first sentence of the book: "If you have health, a great craving for adventure, at least a moderate fortune, and can set your heart on a definite object, which old travelers do not think impracticable, then travel by all means."

According to Galton, adventure was a unique "opportunity for distinction" to the young man who would "probably achieve a reputation that might be envied by wiser men," and one of the "most grateful results of a journey" would be for the young traveler to be admitted into "the society of men with whose names he had long been familiar, and whom he had reverenced as his heroes" (Galton 2000 [1872]: 2). A successful traveler did not hurry, took a passionate interest in his work, had a good temper and knew how to deal with reluctant servants, qualities that would surely have endeared themselves to those promoting imperial expansion, values that Bradburd and Napier allude to. Indeed, "Activating the craving for adventure was essential for progress," Galton later opined in his influential *Hereditary Genius:*

> Luckily there is still room for adventure, and a man who feels the cravings of a roving adventurous spirit to be too strong for resistance may yet find a legitimate outlet for it in the colonies, in the army, or on board ship. But such a spirit is, on the whole, an heirloom that brings more impatient restlessness and beating of the wings against cage-bars, than person of more civilised characters can readily comprehend, and it is directly at war with the more modern portion of our moral natures. If a man be purely a nomad, he has only to be nomadic, and his instinct is satisfied; but no Englishman of the nineteenth century are purely nomadic. The most so among them have also inherited many civilized cravings that are necessarily starved when they become wanderers, in the same way as the wandering instincts are starved when they are settled at home. Consequently their nature has opposite wants, which can never be satisfied except by chance, through some very exceptional turn of circumstances (Galton 1914: 334).

The Art of Travel provides the nitty-gritty of how these explorers/missionaries/traders claimed they went about their business. It was modeled on guidebooks, which were becoming increasingly popular in Europe and which framed the way travelers and tourists saw and described the world they traversed (Withey 1997; Sillitoe 1995). *The Art of Travel* contains surprisingly little advice on how to deal with indigenes in terms of both general etiquette and "extracting information." They were simply part of the backdrop designed to make the explorer-adventurer look good. In general,

Figure 1.1 • How to sleep with a gun according to Galton

and in accordance with prevailing theories, indigenes were seen, ironically, as adults with puerile minds.

Adventurers must not only have to look the part, they must act it as well. In performing their adventures, they took as role models their predecessors who in turn based their performance on the pantomimes of eighteenth-century Europe (Dening 1994). Galton had a keen sense of theater. A sense of the dramatic was important in pacifying potentially obstreperous natives. In dealing with a chief, Jonker Afrikaner, he describes how he dressed in his finest red hunting coat, jackboots, cords, and hunting cap, "a costume unknown in these parts," and defying the "great etiquette in these parts about coming to a strange place" charged with his riding ox right into the door of the chief's house and proceeded to berate the astonished chief in English (Galton 1889: 68–70). In Galton's world, a kierie (walking stick) became a scepter, and a dance became a ball. As Adler points out, "travels performed in a particular manner do not merely reflect views of reality but create and confirm them" (Adler 1989: 1382). Galton's legacy thus very much underwrote and gave credence to the fantasies of the (aspiring) middle class, which was later given further substance in the famous East African Safari of ex-President Teddy Roosevelt, or "Bwana Tumbo" as the locals called him. His huge safari led to numerous newspaper reports and articles as well as another bestseller, *African Game Trails* (1910), and film *Roosevelt in Africa* (1910), which apparently did more to stimulate hunting and tourism to East Africa among the American public than any other single event (Bodry-Saunders 1991: 214–5).

In the Victorian Age, explorer's books exerted an extraordinary influence over the minds and fantasies of youth and adults alike. Adventurer's tales supplied entertainment and drama, and had widespread effects like the formation of working-class "Gordon Clubs" (named after the misbegotten hero of Khartoum), not to mention the later development of various youth movements, including the Boy Scouts and Girl Guides (Moorhead 1960). Their effects were multifaceted. In some cases, by treating nineteenth-century migration as an adventure, they facilitated the heavily institutionalized population redistribution in British dominions (Green 1993). In addition, these youth movements arose partially in response to a significant educational movement in Germany, and later the United Kingdom, which argued that a healthy mind required a healthy body. This also led to various schools like Gordonstoun (attended by Prince Charles), the Outward Bound Movement, and NOLS (National Outdoor Leadership School). All this produced and sustained a cult of sacrifice and heroism in late Victorian and Edwardian Britain, themes discussed by Bradburd, Yengoyan, and Napier. Certainly the most striking exemplar of this cult and the major media event of 1911, the very period when Simmel and Burroughs were writing, concerns the tragedy of "Scott of Antarctica." In retrospect, Scott epitomizes the foolish arrogance of the hero. Scott, like Tarzan, was hampered by middle-class prejudice from studying the transportation methods of the Inuit because they believed that the Will of the White male would triumph over technique (Jones 2004).

In New York, clubs like the *Adventurer's Club* and the *Explorer's Club* were so successful that they formed chapters in other American cities, like Pittsburgh, as Napier notes. In the 1930s and 1940s the *Explorer's Club* would publish books like *Told at the Explorer's Club* (1931) (three printings within three months) and *Through Hell and High Water* (1941). These clubs served as urban havens for the middle and upper class males who ostensibly met to promote expeditions of a quasi-scientific nature (Bradburd calls this "instrumental traveling"). Vetted expeditions were granted the right to fly the Explorer's Club flag, which "is a simple design one that is easily remembered by the natives of distant countries and had been loaned to 107 expeditions in the fifteen years prior to the Second World War" (Explorer's Club 1941).[2]

In the aftermath of the First World War, the new affluence and higher educational standards in the United States meant that as an "educational experience" one could more readily go abroad. Improved ease of global travel and development of technology like cameras provided a major stimulus to "adventure travel," an exercise actively promoted by fledgling national tourist authorities, steamship lines, *National Geographic* and Kodak (O' Barr 1994). The more "off the beaten track," the more "adventurous" it was portrayed, the more status accrual for the participants. The interwar years saw a slew of motorized expeditions, ostensibly engaged in scientific collection but with surprisingly few credentialed scientists accompanying them. The success, and indeed possibility, of these expeditions was fed by the nature of the socioeconomic boom of the 1920s. The myth was common, echoing Galton, that self-motivated persons could make a success of any career, and many of these early Adventurer-Impresarios were typecast in this regard. There are many candidates for the archetype, including filmmakers Martin and Osa Johnson, discussed by Lindstrom. Robert Ripley (of *Believe It or Not* fame) is also representative, as is the traveler-broadcaster Lowell Thomas, friend of presidents and other glitterati (Thomas 1976), who in the interwar years was more famous than any contemporary TV anchor. Finally, the interwar years were the heyday for that unique cult figure, Burton Holmes, "Everyman's tourist," who invented the term travelogue and gave over approximately eight thousand illustrated lectures and grossed over five million dollars for his efforts (Holmes 1977: 11). After the Second World War the tradition continued as variants of Tarzan metastasized. One of these is the Amazonian figure of Jacaré, who, as Neil Whitehead observes, was a key figure in helping shape and channel certain Cold War tensions and fantasies.

There exists a stereotype of the adventurer as the "strong and silent" type, that is, people who do not promote themselves for self-glorification. But adventure's close relationship with visual media complicates this stereotype, and this is one of Tarzan's most important lessons. In fact, visual representation has been crucial in framing public imagination of adventure, and the rise of photography and cinema go hand-in-hand with modern adventuring. It is a well-established truism that pictures by themselves do not create or perpetuate stereotypes. They need to be underwritten by narratives. It is no coincidence that as the machine gun, that tool of mass killing, was being developed and field-tested in the colonies, the camera was spew-

Figure 1.2 • Cover of Valentia Steer's *The Romance of the Cinema.* 1913

ing forth images in the form of postcards. Benedict Anderson has made much of "print capitalism" in the rise of the nation-state in the nineteenth century, but what we are dealing with here is an oft-ignored sequel, the transformation of print capitalism into picture capitalism. One consequence of these technological advances was that the borders between popular entertainment and scientific endeavors became increasingly blurred (Haraway 1989; Mitman 1999).

Not only was the public enthralled by adventure stories, but their imaginations were also fired up by the numerous "People Shows" or *Volkerschauen*. Here the living "savages" would be narrativized by some ostensible "expert." By the beginning of the twentieth century these shows were being displaced, initially by magic lantern and then stereoscopic shows and the illustrated Travelogue, and finally movies. As Valentia Steer observes in her remarkable 1913 book *The Romance of the Cinema*, "There is scarcely a town of any size in the world that has not at least one picture theatre" (1913: 11), and in the U.S. alone there were some sixteen thousand picture theatres at that time, estimated to patronize more than six million people a day. According to Steer, "topicals" (events of the day, such as the visits of royalty and politicians) and travel (especially wars and expeditions) were the main focus of cinema. The camera was and remains an important component of modern adventure. It provides a visual authentication that one has "been there," that indeed, the visual documentarian is very often the hero.

Indeed, perhaps the most ubiquitous form of adventure in the new millennium is *Exhibitionary Adventurism*. It is no accident that the first person to reach the North Pole, Robert Peary, had as a major sponsor the National Geographic Society, and that he took numerous photographs. Or that Captain Scott's famously disastrous expedition to Antarctica produced nearly five miles of film (Steer 1913: 97). The camera, that exemplar of modernity, is important because "its mediating function serves to make the explorer capable of the act of creation itself. Once legitimized in terms of his particular relation to photographic technology, the explorer can claim complete authoritativeness for his vision" (Bloom 1993: 87–88). Given the authority attached to visual evidence in Western society, photographs and especially documentary film footage represent events as authoritative and natural. Houston's father lugged a heavy camera on his trips to the Himalayas. The camera is the immortal eyewitness, documenting both triumph and disaster. Vivanco's examination of the connections between televisual worlds and adventure speaks directly to this legacy.

Postmodern Adventures and (Ad)venture Capitalism

There were other developments as well in the early 1960s, apart from massive time-space compression brought about by improved and more accessible communication and travel. In 1961, Daniel Boorstin notably defined a celebrity as someone who was famous for being famous. The rise of mechanical means of communication and reproduction, and subsequent emergence of media "sciences" like public relations and advertising, had produced a culture of "pseudo-events," events neither genuine nor fake, neither illusory nor real. Manufactured spectacles were designed to generate further manufactured spectacles. The Image had taken over, Boorstin suggested, because Americans could not face ordinary life anymore, in which the excellent and extraordinary were extraordinary and most of the things in life were mundane. Much of what passed for Adventure nowadays, he claimed, was a "pseudo-event." The problem with such a claim, as with the claim of false consciousness, is that it bespeaks an elitism, complete with the irony that it is often the rich and powerful who now can afford to have these "pseudo-events" largely as a status-collecting symbol. The rhetoric of adventure as "pseudo-event" allows for some people to claim the moral high ground, and it can thus be seen as a form of symbolic leveling. Indeed, says Hughes-Hallett, what heroes and adventurers are supposed to do nowadays is dazzle. "The capacity to stage a splendid tableau is a more important qualification for admission to the gallery of heroes than either survival or success" (Holland 2004: 16). This is indeed something that politicians like George W. Bush have realized, as McBride suggests.

Globalization has again transformed the role of our popular hero as his spectral presence is felt in television "reality shows." *Survivor, The Amazing Race,* and *Big Brother* now capture popular attention. Indeed, in Britain, it is estimated that ten million eighteen-to-twenty-five-year-olds voted during the *Big Brother* series, eight times more than voted in the last general election. As Bauman (2002: 61–68) points out, these shows do not simply explain what there is to be explained. More insidiously, they also suggest what viewers should think, and how to think about it. Rather than promote teamwork and community or solidarity as in "old school mountaineering" (per Houston), these shows suggest that one should use the team or larger group for one's own advancement. *Survivor* is about the disposability of humans, and the motto for shows of this ilk is, indeed, "Trust No One." These reality shows encourage lack of commitment, and since trust and faith are the bases of society they are profoundly antisocial. Indeed, Bauman (2002: 72) cites a study that shows how creative energy was based on confidence in oneself and others, and supported by trust in the longevity and undisputed authority of social institutions. With no trust there is no courage to take risks, to assume responsibilities, and to enter long-term relations. Adventurers nowadays have such a degree of emotional and financial security that they can dispense with any need for personal involvement (Bauman 2002: 52). Tarzan or Crocodile Hunter, as Vivanco suggests, does not demand love, loyalty, or devotion.

One of the attractions of adventure has been the camaraderie of the small group. But contemporary adventure tourism, with its clear time constraints and well-structured activities, allows not for *communitas,* but for the creation of what Bauman calls "peg communities," which are formed by the hanging of individual concerns on a common peg, to be a one-day hero (Bauman 2002: 176). Furthermore, survival both in "reality shows," and increasingly in global capitalism, is now predicated on the new keyword: flexibility (Martin 1994). Flexible production procedures create short-term superficial relations at work, and in the local community. In addition, to cope with this short-term flexibility and increased mobility, workplaces and environments have become increasingly standardized, like many so-called "adventure tours," which are supposed to provide a sense of instant achievement. Flexibility and rapid reaction teams are the current fashions. Tarzan swinging from the trees epitomizes mobility, and this is one of the key consequences of the emerging global division of labor that rips asunder ties to place and people.[3]

Furthermore, it is the system of global capitalism that makes it possible for a small segment of the world's population to have the resources to journey afield in order to have ludic adventures. Indeed, one could argue, such is a product of hedonistic affluence. The rich assume a right to travel wherever they want, and do it with a sniff of ethical casuistry and self-indulgence as they wander around poor countries trying to find themselves or "develop" the poor and unfortunate (Iyer 1986). They believe they can escape to places where reality and time are suspended, miracles occur, and they meet the proverbial whore with the golden heart. Adventure tourism thrives on myths and ignorance. In this sense, adventure serves to display the adventurer's economic and cultural power. For many of the locals, the Adventure is a form of conspicuous consumption that exhibits the adventurer's powers of acquisition and taste. Indeed, the economic power of the adventurers remains at the center of the threads that link adventure and sex (Littlewood 2001: 93). Even the feminists who demonstrate their sexual independence by taking black lovers are at the same time rebels against dominant values in their own society and exploiters in the host society, as Mathers and Hubbard show. One can be a rebel against dominant values and exploiter of them at the same time.

Hans Magnus Enzensberger has pointed out that nowadays the global elite see it as a human right to distance themselves as far as possible from their own "civilization." Paradoxically, the destination has to be both accessible and inaccessible, distant from civilization yet comfortable, dangerous yet safe (Enzensberger 1997: 127). It is hardly ever an adventure in Simmel's sense because if things got bad they could leave: "As we point to the return ticket in our pockets, we are admitting that freedom is not our goal and that indeed we have already forgotten what freedom is" (Enzensberger 1997: 135).

Ironically perhaps the person who has done more to undermine contemporary Western Adventure and contribute to its *ersatzization* is Galton. Among his many innovations and discoveries was his seminal contribution to the burgeoning industry of "risk management." He did this by "transforming the notion of probability from a static concept based on randomness and the Law of Large Numbers into a dynamic process in which the successors to the outliers were predestined to join the crowd at the center…. Regression to the mean motivates almost every variety of risk-taking and forecasting." This was to form the basis of the actuarial tables used in the insurance industry to insure against risk (Bernstein 1998: 170). According to the U.S. Census Web site, in 1997 the insurance industry was worth over $59 trillion and there were more than 352,603 licensed insurance agents. By 2002 the value of the industry had almost doubled to $105 trillion. With such insurance coverage, does any member of the privileged classes take an adventurous risk?

Anthropologists and/of Adventure

And what about anthropology? Anthropology has long had a romance with adventure. There is little doubt that its public identity is intertwined with it, a fact relevant long before the likes of Indiana Jones came along (as Yengoyan's essay here shows). Franz Boas himself was drawn to anthropology through its association with adventure: his favorite books were *Robinson Crusoe* and Humboldt's *Cosmos,* and he practiced eating foods he didn't like, "In order to accustom myself to deprivations in Africa" (Pierpont 2004: 51). Many contemporary anthropologists will admit that it was a similar desire for adventure that attracted them to the discipline, a sensibility very much framed by mass media images.

But the association with adventure and adventurers has also been cause for tension within the discipline, as demonstrated in *Tristes Tropiques,* Levi-Strauss's most famous book, which opens with the words, "I hate traveling and explorers. Adventure has no place in the anthropologist's profession." Of course, in denial he emphasizes its place, reflecting the contradictions in the discipline's efforts at differentiating itself from Euro-american others (missionaries, explorers, traders, administrators, travelers, etc.), those figures often present in fieldwork but muted in traditional ethnographic writing and monographs. This is perhaps what makes adventure such an intriguing and productive theme for thinking about and contextualizing anthropology itself, for it means exploring the discipline's emergence out of the same historical and political contexts that produced adventure-seeking sensibilities and the visual spectacles associated with knowing otherness. It also entails an examination of how the discipline's own conventions of professional fieldwork and ethical praxis have been shaped in relation to (often in differentiation from) adventurers and adventuresome experiences, issues which Stoll's chapter in this volume directly addresses.

Adventure is also a rightful topic of anthropological inquiry, as a cross-cultural phenomenon. Can shamanism, which is intimately connected with the notion of a break with the mundane and an epic and highly hazardous journey, be conceptualized as adventure? Rubenstein addresses such a question, showing a sharp contrast between Shuar and Western concepts of adventure, and arguing that any understanding of non-Western concepts of adventure must begin with appreciation for very different phenomenological precepts and social practices. There are broader lessons here for a cross-culturally nuanced anthropology of adventure. Indeed, in many parts of the world, including Africa, migrant labor is treated as an Adventure (Luig 1996), but the goal is not necessarily self-discovery or self-enrichment. Given that African definitions of the self are not of the autonomous billiard ball variety but rather believe the self to be constructed out of myriad social relations and obligations (Piot 2002), a different version of adventure might be expected. It is suggestive that Rambo is popular in many parts of

Africa and the Third World, not because he is a tough adventurer, but because he is seen as fulfilling (pseudo) kinship obligations and fighting an intransigent and corrupt bureaucracy.

This volume covers the range of these concerns and the distinct kinds of research and analytical approaches they imply. As anthropologists we are convinced that one of the important roles of anthropology in the new millennium is to serve as tricksters and as intellectual *agents provocateurs,* both within and outside the discipline. Regarding the former, this volume is transdisciplinary in its methods and insights, drawing on insights of and in dialogue with disciplines like history, philosophy, literary criticism, film studies, political science, and gender studies. Yet, regarding the latter, this collection aims to do what any good anthropology should do, that is, both teach us something about what people are doing in the world and shake up our most common and taken for granted assumptions about what adventure is and what adventurers do.

Notes

1. In this regard Paul Theroux's essay "Tarzan was an Expatriate" (1998 [1967]) was inspirational.

2. Again the gender dimension is significant. In 1925 four women founded the Society of Women Explorers to encourage and support women engaged in such activities since no "explorer" organization would allow female members. This society, *The Society for Women Geographers,* still exists today with more than six hundred members, still however small change compared to the dominant men's organizations. The National Portrait Gallery has helped to correct this bias with a well-received recent show, "Off the Beaten Track: Three Centuries of Women Travelers" (Riding 2004).

3. Space precludes a discussion of the role of adventure as ideology in the development of capitalism. Suffice to refer the reader to Michael Nerlich's (1987) two-volume classic on this topic in which Weber's famous thesis on the Protestant Ethic is given a thorough airing.

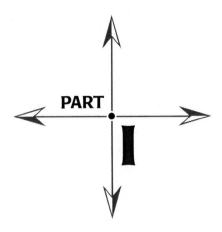

The Adventurous Worlds
of Simmel and Tarzan

CHAPTER 2

Simmel and Frazer:
The Adventure and The Adventurer

Aram A. Yengoyan

Adventure, by its various definitions and by what it embraces, comes in many expressions, some might be near as an aspect of quotidian life, others might be far and distant such as dreams and venturing into totally new contexts. Adventure and anthropology are coupled like Siamese twins. The reasons are numerous and they take us in many directions, thus, I will explore what I consider some of the important ones. The best of adventure might tell us something different and new from normal anthropological enquiry and even from abstract theories like structuralism. Anthropology was born out of adventure, and its roots are quite recent in comparison to how the West understood the foundations of adventure and exploration, which were prior to anthropology's nineteenth century foundations. Since anthropology started as an academic discipline, its empirical findings came from adventure and adventurers in the form of colonial officials, explorers, missionaries, administrators, and the intrepid traveler. One can quickly note that, in the broader sense, mutatis mutandis, the necessities of forging anthropology as a science could only have been done by denigrating its roots in adventure. Examination, however, of Georg Simmel's and Sir James Frazer's writings on adventure reveal the basic juxtaposition of adventure and anthropology.

There is a newness to adventure, and of course the concern is how does one interpret this newness and what might it mean. Yet, we commonly associate adventure with various forms of modernity, a modernity that might deal with forms of global transformation or one that deals only with the realm of psychological differentiation, be it Freudian or any other type of psychological venturism. The realm of adventure that was explored by Simmel and enacted through the voluminous writings of James Frazer differs in regard to what modernity meant, but in both writers there was a strong sense of lamentation that converged at various times.

Modernity is a form of thought progression that links enlightenments through time and space. Each phase of these enlightenments is historically grounded, yet in the analysis of the evolution and the trajectory of these enlightenments, the presuppositions of the foundations of each grounding and each source must be questioned. Throughout Europe from the 1500s onward, different sets of enlightenments expressed different, yet overlapping, modernities regarding how they were grounded historically in the past, and also how modernity dealt with utopianism through national historical traditions. Thus the English, French, German, and Scottish enlightenments differed in many significant ways, yet in each case modernity, even a nuanced modernity, was critical to how that ground was established.

However, the ground of each enlightenment was examined and reexamined from different sources. Through a reexamination of the foundations of history and thought in the context of enlightenments, one may ascertain how various degrees of uncertainty emerge, why they persist, how they are reworked, and finally the extent to which they underwrite modernities. As elaborated by Nerlich (1987), adventure itself is an epic of modernity, but as he brilliantly shows, adventure has its roots in a period in European thought and history that was precapitalistic and also premodern. Adventure is something new but like all newness it becomes routinized, which in turn, dialectically calls for other adventures which are new or a new reforging of bits and pieces of the past.

Modernity challenged ideas of economy, polity, history, and the basis of legitimation, science, and of course, religion and metaphysical thought. Except for religion, and possibly metaphysics, postmodernism has in turn challenged the very foundations of what modernity wrought. However, most of the treatises on postmodernism seldom tackle or severely question modernism's critique of religion be it an experiential state or an institutional form.

In exploring why this is the case, one can only conclude that for what modernistic philosophers and thinkers like Kant and eventually Freud understood as religion, it was only a matter of time until science and psychology would uncouple religion from metaphysics, thus in turn opening the ground for reason and only pure reason. An example of this uncoupling is

the idea of the soul, which from the ancient Greeks onward was always understood in conjunction with memory. However, from the 1850s to the 1910s,
memory and the soul were detached from one another through French
medical practice and the emerging science of psychology (Yengoyan 2004).

Furthermore, one could also argue that the kinds of psychological liberation that underwrite postmodernism can be acquired and sustained through
religious and metaphysical experiences that cannot be institutionalized or
legitimated through the structure of statelike apparatuses. In a peculiar way,
Simmel and Frazer are relatively similar in this regard. Simmel's writings
on religion are very psychological, while Frazer saw magic and metaphysics
as an appeal to certain psychological states that simply cannot be reduced
to the scientizing of psychology. In more recent times, the attractiveness of
the writings of Carlos Casteñada also bears out this appeal to another life
form, but now the new coherent message is between religion, shamanism
and hallucinations that could be drug-induced. Surely we cannot assume
that the role of religion and the reemergence of metaphysical thought were
laid to rest by the enlightenment(s). Simmel realized religion would be in
conflict with a secularized modernity that hardly possessed the authority
and the ability to challenge the tenets of religious thought during his time.

For Frazer, the appeal of magic and religion was that life was full of sur-
prises that could not be deduced from the material circumstances of any
society. This, of course, was also a dominant concern for Durkheim in his
insistence that religion, meaning and ritual formed a logical entity that
could be comprehended without resorting to mystification. Yet, Frazer
went further in his insistence that these mystical differences ranging from
illiterate cultures to complex societies were a living testimony to the creativity of the human mind as well as the open-mindedness of what was humanly possible.

However, both Simmel and Frazer were also very Kantian, especially in
assuming that nature is the great constant and the quality of nature as being the same from one context to another. By assuming this position, both
writers stressed that humans can do more, which in essence is one version
of true Kantianism. For Simmel, the ability to do more was partly expressed
in his ideas on adventure, namely the ability of individuals to go beyond
the mundane quotidian aspects of life. For Frazer, the ability to do more was
partly expressed in the ongoing mystery of life and thought, the venturing
into new and different arenas of magic and religion that cannot be predicted
from one context to another.

Overall, Frazer concludes that there is a profound and deeply felt admiration for magic and religion and these must be understood as parallel to our ideas of devotion to science. Metaphysics has an equal place with science, though Frazer realized that the thought of the enlightenment(s) would work against what he saw as human creativity and the ability to venture into new forms of thought. In this sense the tragedy of modern life for Simmel was modernity and its impact on the human psyche; for Frazer the tragedy was a form of loss of those humanly created endeavors that no longer had an expression in the scientism of our times (Yengoyan 2002).

Both Simmel and Frazer are dedicated to the idea of the universalization of humans, a growing sense of humanity and how this humanity is expressed. For Simmel, this universalization of humans was demarcated by the dilemmas and paradoxes that impacted on individuals and institutions when traditional cultural structures were uprooted as meaningful for the actions and thought of individuals. Much of Simmel's writings on the general human conditions of his time are characterized as being pessimistic, morose, and even fatalistic in that there is no exit from this general state of being. Frazer also laments the passing of the world as a whole and in many directions his final words on the collapse of what he envisioned echo the melancholy and dejection one finds in Freud's later works.

Yet, Simmel and Frazer represent points of departure on the ongoing debate between religion and secularization. Each writer saw the problem through a different lens, though the issues were not that far apart. Religion during Frazer's time was changing and has changed more so in the past five decades. If Frazer lamented the progression of change from magic to religion to science, most of the New Age writings on religion attempt to reverse this process by focusing on the shift from science to religion to magic. However, the progression of these shifts still cannot account for the fact that verbs that describe religious states are in decline and that probably started even when Frazer was writing. These shifts are difficult to fully comprehend but I think that the reduction of religious verbs in English has been accompanied by an increase in verbs that refer to psychological states, either in normal usage or in its abnormal counterpart such as psychiatric states or in parapsychology.

Secularization, which for Simmel was one component of modernity, must also be addressed and we might ask if secularization has passed or is passing. Both Simmel and Frazer saw its persistence and Simmel fatalistically felt that it was a state in which we could not escape. Frazer also understood this process but his insistence on the wholism of the human experience must be fully realized against processes of modernity and secularization that fracture and dissolve the past. Thus, the deadening effect of these processes not only annuls and disintegrates the past, it also levels and flattens the quality of social life into innumerable cultural and materialistic items and facts that could not be internalized by the individual. This is a dominant theme in Simmel's writings on culture and modernity. In Frazer, the death of the past and what it embraced in terms of differences and uniqueness was a process he lamented and yet tried to capture in the twelve volumes of *The Golden Bough*.

Here the similarities and contrasts between Simmel and Frazer are developed along the lines of adventure. For both writers adventure was a journey that brought forth new developments, new vistas that one could not predict or understand. In this sense adventure was a break, possibly a radical departure from what is known, what is mundane, and a departure from the constraints and restrictions of a politically induced social milieu that curtailed the aspirations of the individual. Simmel and Frazer might have meant different things by adventure, yet there was a certain common linkage central to their thinking. The centrality is based on what each writer conceptualized, envisioned, and expressed by adventure and how the idea could be conveyed.

Simmel, the essayist, explored various facets of how psychological identity could be established within and apart from the conventions of his time. Conceptions of the self for Simmel could only be attained to a certain and limited degree through what formal sociology had to offer. Not only did formal sociology constrain what Simmel was apprehending, in most cases it asked questions that curtailed and limited the kinds of freedoms and ventures that Simmel sought as basic to the individual. Much to the chagrin of Durkheim, Simmel through his essays attempted to isolate the impact of modernity and in turn to offer various pathways by which the individual could achieve a sense of selfhood and excitement beyond the ongoing constrictiveness of social life.

For Frazer, a sense of adventure and the various expressions of that idea could take place through a number of related means. One was that Frazer saw his venture as a literary quality, the ability to write about past times and past events. Through the overwhelming use of customs, cultural traits, beliefs, ethnographic facts and fictions, and odd and unique thought, Frazer saw this breadth of humanity as a form of adventure that could be understood and appreciated in the coziness of one's home. As a literary personage and as a humanist, Frazer also lamented the impact of modernity through global imperialism, which altered and destroyed the very subject matter that he was transmitting to an eager public audience.

Simmel and the Adventure

Simmel's (Frisby and Featherstone 1997) most insightful essay on the adventure was published in 1911, yet it is still one of his most perplexing statements, which can be read and interpreted in numerous ways. Apart from the topic itself, the whole of Simmel's cultural philosophy and sociology is captured in the brilliance of his method of expository writing.

Adventure embraces the issues of distance and differentiation. Distance is created through marked departures regarding space, time, sexuality, and eventually forms of representations, all of which engulf the individual within everyday quotidian structures that are hostile to experiences that are new, unique, unconnected, and possibly unexplainable. Once adventure is created, it must be sustained through a sense of alienation from ordinariness, from history, from the past, from the future, and from a bounded structured existence.

Distance and differentiation are interconnected, yet the idea of difference is critical and essential to breaking the synthesis of organic bonds that created the deadening dominance of the mundane. The creation of the non-organic curtails the safety of contingencies; it moves action and thought from certainty to uncertainty, and eventually a move to openness. Yet, Simmel's message is one that argues that differentiation is the most pivotal point and concern through which adventure can be realized. One theme of difference is that proximity breeds or yields inversion, thus the issue of opposition can occur at all levels of action and thought. As Simmel notes, the adventure, be it near or far, is a break and a critical departure from what is deemed as social or as society. Thus, exoticism, eroticism, and dreams are all states of action and mind that represent a rupture from the essentialism that society creates and demands as acceptable.

The ahistorical persona is able to create distance from the past, to forego the future, and to maintain a context that has no beginning and no end. Escape and distancing is fundamentally an issue of structure and antistructure that has been central to the writings of Victor Turner (1967, 1969, 1972).

However, Simmel meant more than the structure/antistructure dichotomy. What is critical for Simmel is how the individual can break and depart from social forms in a way in which adventure is created as a means of finding oneself.

Adventure is based on a rupture, a form of isolation, a separation of self from the whole and the creation of accident(s) that are apart from the continuity of life. In this vein, risk, danger, chance, and all accidents lie outside of continuity and thus chance is always jeopardized by the movement toward necessity that maintains continuity. Freedom and the adventure are not simply a matter of the individual and social form, but eventually a means in which the mind and knowledge escape and are released from the limiting rational frameworks that embrace and dictate our totalizing minds and methods.

In Simmel's time, the only philosopher who keenly understood what was at stake was Nietzsche. Through sarcasm and wit, Nietzsche's *Thus Spake Zarathustra* is a clarion call on the evils of the state as the new idol that leads to the slow suicide of life. True intellectual achievement and creativity is only at the margins, those spaces the state has yet to penetrate and control. For Simmel, like Nietzsche, the collectivity of the state and the social fulfills only the lowest levels of human intellectual achievement (Axelrod 1979).

For Nietzsche, the foundations and sources of human creativity exist only in the margins of society that are far from the tentacles of state domination. Simmel also realized the deadening quality of the social and the collective. The only means of transcending the social was through fragmentation, the ability to create tissues and bonds between individual and institutions that were contingent and not binding. Fragmentation is thus a situation of fewer and fewer threads, a context that creates the conditions for adventure. Ordinary experiences are totally blanketed by social ties that promote continuity, form, boundedness, and anti-isolation. Thus, chance in society must be controlled and minimized, but for the adventure, chance is a wholly positive condition.

Fragmentation is essential for creating the antisocial context that is the basis of adventure. Yet, adventure is not simply the reversal of the social and adventure is not there to critique the social. What Simmel is asking is how social forms can be altered to create context(s) that are fragmentary and through which adventure may be enhanced. In his time, the routinization of German society through the Prussian state and political structures curtailed and denied any sense of adventure.

If the tearing of life completely out of itself was the task at hand, so be it. A social form that has no space for voyeurism, which is an adventure within itself, must be critiqued and transformed, it must be made responsive to basic individual wishes. One can read Simmel on the adventure as a personal statement, but also adventure is the only form of escape one might possess. Modernity and how it evolved in Europe after the 1870s was hardly a sense of escape from Wilhelmine Germany. In fact, the lasting impact of modernity was barely felt in reforming the political structures that embraced Europe in the pre-1914 period. Modernity was accepted as a veneer that national polities utilized and portrayed to society; but modernity itself was never organically interconnected within the national polity.

The estrangement between the group and the individual is expressed in a number of ways, and these expressions also change through time. In Simmel's time, the movement from Imperial Germany and the veneer of modernities created one crisis after another. Disenchantment, as a reigning concern throughout Europe combined with various degrees of estrangement within the social fabric, is the context that promoted fragmentation. Thus in Simmel's case, the fragmentation of his thought through his writings must be partly understood through his relativistic approach to social and cultural phenomena. Fragmentation is hardly a negative quality; it is basically another means of describing social and cultural complexities and a means of creating different readings on this array of complexity. Each of Simmel's essays renders a different reading with the aim of understanding the infinite variety of meanings and interpretations that may emerge.

Returning to Simmel's idea on the adventure and what it means to the individual brings forth another set of concerns. If the adventure is ideally a radical break, an acute departure from the past, from the future, from history, and from structure, my concern is, how does the individual who is in this state possess the ability to understand what is happening? The question we must ask is, what does one need to know for knowing and understanding something? I find it difficult to detect in Simmel's writings how radical and unknown experiences as adventure, be it erotic or nonerotic, can be understood if one does not possess the *prior text* to understand these events.

The *prior text* is what each of us possesses to comprehend and internalize events. Texts are based on prior texts, which is a form of language (Yengoyan 2003: 25–43) that underwrites the various texts that we live in and through. We all possess prior texts but these partly vary from one individual to another. A good personal example is, what are my prior texts for understanding popular music? Thus, punk, rock, rap, reggae, hip-hop, ska, heavy metal, Christian heavy metal, and the Sex Pistols are all noise to me and I cannot distinguish one from another. I do not possess the prior texts to make sense of this music, let alone to demarcate one from another. Yet, my children have the ability to understand each form, how to internalize it and how to designate markers that differentiate one tradition from another. They possess the necessary prior text(s) to understand these musical forms, which are totally foreign to me.

If I am reading Simmel correctly or within his perspective, one must ask not only how adventure is created and enhanced, but how it can be understood, and eventually, how it is internalized. Surely Simmel's sense of adventure is more than tourism or an Alpine journey. Adventure must be based on distance, alienation, a denial of the present as structure, and a denial of the past, thus true adventure must be ahistorical. Given the kinds of experiences and situations that are the basis of adventure and a radical release from the quotidian, the question that must be asked still remains. Experiences, both internal and external, which we cannot comprehend and/or internalize remain outside of the individual. The essentialness of the *prior text* is its ability to move experience to a sense of knowing.

Frazer as the Adventurer

Sir James Frazer's *The Golden Bough* in either one, two, or twelve volumes was written over a span of time starting in 1890 and culminating with volume thirteen in 1936, just five years prior to his death. From its inception, the anthropological, historical, and literary responses to the work have been massive and now, virtually over one hundred years later, the responses and discussion are still carried forward with varying degrees of intensity.

My major concern in dealing with *The Golden Bough* is to ask why the whole enterprise became so critical to the reading public as well as to academics. There are numerous factors that are vital to understanding its popularity, and over the past hundred years the reasons for the exceedingly high and increasing book sales have partly changed, while other reasons have remained more or less constant. Furthermore, how *The Golden Bough* has been employed and accepted in recent times not only is partly explainable, but in most cases their connections to Frazer would have been difficult to predict.

If Simmel dealt with adventure in a highly original and sophisticated style especially in setting forth the psychological dimensions, Frazer was the armchair adventurer. It was not a matter of psychology, but the overall concern was to dwell on facts and strange customs and differences, with all of these differences still reaffirming Frazer's conviction of the homogeneity of the human mind. It was a total odyssey that started in the United Kingdom to Nemi, spanned the globe as Frazer knew it, and returned to the comfort of one's home.

The idea of the golden bough appears in many contexts prior to and during Frazer's writings. In 1834, J.M.W. Turner (1775–1851) painted *The Golden Bough* as the sacred grove at Lake Nemi (Frazer thought it was Lake Nemi, though it was probably Lake Avernus), home to a critical yet minor sect in Roman mythology. However, in Turner's painting, the Sibyl is holding the branch, thus it is difficult to determine or interpret if the Sibyl is meant to represent Diana, the queen of the woods. And in 1927, William Butler Yeats used the golden bough image in his poem "Sailing to Byzantium."

As an encyclopedic venture, Frazer was able to use historical and ethnographic data as a literary task at hand, thus most of the writers on Frazer would follow Leach in stressing that Frazer's contribution to literature was the issue since *The Golden Bough* hardly contributed to science. Furthermore, Frazer was attempting to tap into what might be called "mental anthropology" or the investigation on how individuals and people think. But for Frazer, the mental life of people is in the writing of literature and not in history or science. Thus the evidence, the raw observations of peoples and beliefs all became the material for superb writing and stylistics.

In the past four decades, Leach (1961, 1966, 2000) has been the most vocal critic of Frazer while other voices coming from literary backgrounds such as Weisinger (1961) and Manganaro (1992) are more supportive or at least open to the Frazerian venture. The debates also involve theology, comparative religion scholars, and historians focusing on postcolonial studies.

Yet the popularity of Frazer has not waned and the sales of the volumes also indicate this trend. Ackerman (1987: 256–57) notes that Frazer received an extraordinary 25 percent royalty in the 1920s on the third edition, a contract that was unheard of in those days. Between 1911 and 1922, thirty-six thousand copies were printed and sold. The twelve-volume version remained in print into the 1920s and each volume was printed two or three times separately during that decade and into the 1930s. Furthermore, in

1977 Macmillan again reprinted the twelve-volume edition. And Beard (1992: 213) also notes that sales had not waned as of the early 1990s. Toward the end of this section, the matter of sales will again be discussed.

Still we must ask, what did the reading public see in the work, why was it bought and read or purchased as gifts? Many English households had copies of *The Golden Bough* just as numerous American households had lazy Susans. Furthermore, as Beard notes, *The Golden Bough* was a gift that might not be read, but it would fit into a bookshelf, just as Americans kept but did not use lazy Susans.

Beard's (1992: 203–24) essay on the popularity issues covers a broad range of subjects that were critical to understanding sales. As an odyssey or even possibly as a sense of voyeurism, *The Golden Bough* appealed to a middle and upper class reading public for a number of reasons. The volumes are of a gifted literary quality and the prose is skillfully crafted and brilliant. Furthermore, it had answers to explain bizarre customs, it could be read for popular knowledge of common problems such as the mother-in-law taboo, it had close connections to exploration and British imperialism, and it was also a useful and practical reference book. On more abstract matters, it could be read as the legitimation of colonialism and it assured the reader that British imperialism could be of value; it showed how savage custom related to the ancient world and to the present, and that savages could be incorporated into the ancient world and into contemporary Britain. And finally, the comparative project was geared to the reading public and that people—past and present—could be understood within a single intellectual program. As a tale of travel and exploration, *The Golden Bough* was a metaphysical voyage from the unknown to our world, with Nemi as the start and as the end.

Still, Leach emphatically feels that after 1900 Frazer said nothing new and eventually he was only a popularizer and propagandist. This opinion was partly shared by Jane Harrison, who was a colleague of Frazer's in the 1920s. Yet, the mystique was there in the early 1900s and is still with us, though somewhat tarnished. On the literary side of Frazer, some critics feel that his appeal in *The Golden Bough* was that it presented the issues as a form of romance in a highly gossamer style that attracted the reading public.

Others feel that Frazer's work had a direct appeal to missionaries and religious people by dealing with the ideas of evolutionary process and how they related to the trajectory of magic to religion and eventually to science.

The ideal opening in *The Golden Bough* was Frazer's ability to find modern parallels between Christianity and other religions, but more so with Middle East cults, which were critical for the understanding of the Old Testament.

As one would expect, reading Frazer would also partly appeal to the sexual interests of Victorians in the United Kingdom or elsewhere. Various vignettes of sexual custom, organ mutilation, etc. are noted but the heavy kinds of graphics one might desire then and now are absent. Leach (1966) claims that the pleasures of reading about sex from Frazer fit into a sort of "sado-masochistic sexuality," mostly primitive sexuality that veers to polite pornography that the Victorians could accept.

Anthropology's pornographic agenda might have been in the closet in the nineteenth century, but surely by the time that Malinowski, Mead, and Frazer were writing from the 1910s to the 1930s, the secret was out of the closet. Sexuality might probably have sold many copies prior to the 1940s, but I suspect that this issue is no longer a driving force in regard to the current sales of *The Golden Bough*.

Frazer Anew

Over the past two to three decades, the Frazerian adventure has created new vistas and insights, which are again reflected in the sales of *The Golden Bough*. Within the past ten years, *The Golden Bough* has averaged about five to eight thousand copies sold per year. One way to assess its current popularity is to examine its sales on Amazon.com. Actual sales on Amazon are difficult to assess, since they are ranked from 1, which is the highest-selling title on the site, to over 3,000,000, which is the lowest. Thus, Amazon provides only a ranking relative to all other titles they offer and not actual sales, and these rankings change daily. For *The Golden Bough*, Amazon lists seventeen different editions in print, one of which is the thirteen-volume library edition. The most popular edition is a current paperback. This was recently ranked at 25,688 and it had twenty-one volunteered customer reviews. For a work of ethnology, this is truly exceptional. By contrast, Clifford Geertz's *Interpretation of Cultures*, certainly one of the best-selling titles, ranked 27,725 and had only three reviews. Unlike Frazer's *The Golden Bough*, Geertz's work is frequently assigned in universities. Frazer's support is mostly extra-academic. Moreover, the reviews of *The Golden Bough* are extremely positive and supportive, something Frazer seldom experienced during his life.

Renewed interest in Frazer's work, which is the basis of these sales figures, comes from a range of religious movements such as supernaturalism, mysticism, new religions, and neopaganism in its many variants. Reading lists for pagan groups, the study of witchcraft, new-age Siberian shamanism, occult philosophies, Druidism, the Hermetic Order of the Golden Dawn, new spirituality, Enochian Magick, mysticism, and the Mother Goddess movements all list *The Golden Bough* as required reading and in some cases the two-volume edition is mandatory.

Furthermore, Wicca, the most important of these pagan/neopagan religions, was founded by Gerald Gardner, who was heavily influenced by Frazer and his works, so much so that some have argued that in order to understand the origins of modern Wicca, Frazer must be initially read. Gardner created and popularized Wicca in the early 1900s and was its major intellectual force until his death in 1964. Yet, there is a critical debate over what Gardner borrowed from Frazer, how much Gardner fabricated, and what are the roots of modern Wicca. However, *The Golden Bough* resonates loud and clear in Gardner's *Witchcraft Today*.

Within the United States, Frazer's writings are critical to Margot Adler's *Drawing Down the Moon: Witches, Druids, Goddess-Worshippers, and other Pagans in America Today*, 1979. It is a well-respected volume that contains numerous references to *The Golden Bough* throughout the text. As the first of the pagan and neopagan works published in this country, it is now considered a classic by most followers. Probably the best statement regarding the critical and lasting impact of Frazer's writings on modern pagan movements and pagan witchcraft is to be found in *The Triumph of the Moon*, by Ronald Hutton, 1999.

Frazer's work has undoubtedly been used as an intellectual support and justification for the revitalization and emergence of various pagan practices. Sales of *The Golden Bough* in North America, Europe, and Australia all indicate this renewed interest in Frazer. It is no longer read for its sexuality, but primarily to understand the savage within us. Frazer's intellectual voyeurism has no limitations.

But *The Golden Bough* is not limited to the foundations of modern neopagan and occult movements. In 1979, the film *Apocalypse Now*, directed by Francis Ford Coppola and starring Marlon Brando, was released. The film is a loose remake of Conrad's *Heart of Darkness* within the context of the Vietnam War. Brando, who is cast as Colonel Walter Kurtz, an unhinged officer right out of Conrad, is sitting and waiting for the end to arrive. Kurtz is reading T.S. Eliot's *The Hollow Men* and on his desk we find a copy of Sir James Frazer's *The Golden Bough*, the ritual sacrifice of the god/king Kurtz lifted straight from the pages of the mythology compendium.

In Defense of Adventure

In *Tristes Tropiques,* Claude Lévi-Strauss's opening paragraph is a firm indictment of the role of adventure and the adventurer in anthropology. As he notes, adventure is a drawback, a negative one, which we have been involved in, but it has no place in the profession of anthropology and in the work of the anthropologist. It is a hindrance like so many other things such as sickness, exhaustion, and hunger. Whose hunger we are not told.

Yet, while Simmel and Frazer differ in many ways, we must conclude that their form of adventure and adventurism has had a place in anthropology in both earlier phases and the present, although the issue is to deal with Lévi-Strauss and his ready dismissal of adventure and adventurism. It is true that much of adventure writing is of a farraginous quality and to a limited degree we also find this in Frazer, but Frazer was also a highly imaginative writer, one who achieved a pinnacle in literary belletristic exposition.

Apart from writing, however, adventure and how it was understood was a theoretical expression for Simmel. For Frazer the serialization of fact and fiction might be described as a form of hebetude if taken in large doses in one sitting. The reason why so many readers find Frazer dull is that the lengthy periods of reading produce a sense of solace that is difficult to dispel.

In opposition to Lévi-Strauss's views, notions of adventure have played an essential role in the shaping of anthropological thought and interpretation. For instance, for Levi-Strauss, the matter of travel and adventure that he (1973: 17) "hates" becomes a past that must be erased in regard to the theoretical and intellectual foundations of the field. Not only was adventure/travel prior to anthropology, it virtually provided the empirical foundations for the field prior to the idea of systematic anthropological fieldwork, which emerged with Boas.

Second is the continuing interest in the use of travel accounts, adventure literature, and histories in the context of anthropological explanations and understanding. Reading and utilizing Herodotus for an understanding of the ancients is basically the use of adventure literature, though the work is considered a form of history. Virtually all anthropological fieldworkers must rely on accounts written by travelers and adventurers, however we all know that such sources vary in quality, the accuracy of their observations, and the kinds of prejudices which are overt and covert in the text. Social anthropology has its roots in such observations and over time we have gained the ability of knowing how to use accounts and texts with a greater degree of exactitude.

Lastly, we are all involved in adventure. This might not be what Simmel and Frazer meant, but the idea of adventure that I am proposing operates on hunches, the concern to be involved in speculation, and also realizing that our normal kinds of enquiries can only take us so far. This is what fundamentally occurs in fieldwork when things are "discovered" that cannot be accounted for through our normal theoretical and methodological inquiries. Since cultural logic is local and specific and not a matter of pragmatics, we would expect certain societies to have the ability to intellectualize certain features of language and culture in ways that are quite complex and highly variable.

This process of play or involution is a form of cultural creativity that is usually nonpredictable and commonly done through language games. Anthropology does not have a framework to understand these games and we are never sure when these games will or will not occur. But most fieldworkers, especially those working in the western desert of Australia, have encountered cases of cultural play; the best we can do is to speculate and describe what we have noted, and that is usually the result of a probing mind. Venturing into the unknown and unpredictable occurs when one's disciplinary foundations have run their course. Adventure is an imperative, a mutatis mutandis, which we must follow when our normal inquiries cease or are no longer responsive to what we have encountered.

Anthropology was born out of uncertainty, and to this day the issue of uncertainty must remain paramount. One might respond to this by claiming that our methods and theories explain uncertainty, and this might be the case but to a very limited degree. Our efforts are hardly scientific, our theories are fleeting at best, and our subject matter is always a matter of debate. All of our concerns regarding chance and situationalism support the claim that not only are the foundations of anthropology transitory, it is this openness, created by an uncertainty of the subject matter as well as the empirical evidence, which has continuously provoked interest in anthropology.

Creativity, based on chance and uncertainty, is part of our general human condition. Just as Simmel argued against social and temporal forms that limited adventure, his position can be extended to the claim that anarchy, chance, chaos, disorder, and antistructure are all an inalienable part of our reality. The realm of uncertainty is not only what makes life creative and venturesome, but without it the social and political realities as we know them will remain binding and sterile. New orders and structures are forged and created out of adventure, chance, and uncertainty, but as these new orders become more predictive and routinized, a new phase must ensue that renews and promotes a dialectic in which chance and adventure are again on the ascendancy.

Adventure, as an odyssey and as voyeurism, is never limited to the physical force of bodies and selves encountering what is new and uncertain. The odyssey of the mind must be paramount since what develops as mental puzzles and questions will be the most lasting. A mind that seeks chance and uncertainty is the very basis of what Simmel was appealing for as a general human condition.

But why does Lévi-Strauss hate traveling and explorers and why does he feel that adventure has no place in the profession of anthropology? Is it an issue only for Lévi-Strauss or is it a general French condition? Nerlich (1987: xvii) quotes for us:

> ... We would like to see a philosophical history of adventure. What tendencies and fantasies contributed to adventure at certain times? How was adventure set in motion? How did adventure erupt as it did?—How, according to the rules of the imagination and the circumstances of history, were ideas such as love, religion, virtue, and chastity able to interweave themselves so as to make adventure, and how did the proportion of this or that idea change through the intermixing of peoples and the transformations of time? How did France, for instance, lose its taste for adventure? Where and why can remnants still be found in Europe?
>
> Johann Gottfried Herder

Notes

I wish to thank Dana Herrera, Kendall House, and Michael Winter for their assistance on various phases of this paper.

CHAPTER 3

Adventure in the Zeitgeist, Adventures in Reality: Simmel, Tarzan, and Beyond

Daniel Bradburd

Introduction

As it recounts Tarzan's feelings on his first return to the jungle after he has acquired a veneer of civilization, Edgar Rice Burroughs's *Tarzan of the Apes* highlights civilization's constraints. Challenged about his knowledge of lions by fellow Europeans, Tarzan accepts a wager to hunt a lion, naked and armed only with a knife and piece of rope.

> *Tarzan had no sooner entered the jungle than he took to the trees, and it was with a feeling of exultant freedom that he jumped once more through the forest branches.*
>
> *This was the life! Ah, how he loved it! Civilization held nothing like this in its narrow and circumscribed sphere, hemmed in by restrictions and conventionalities. Even clothes were a hindrance and a nuisance.*
>
> *At last he was free. He had not realized what a prisoner he had been.*
>
> *How easy it would be to circle back to the coast, and then make toward the south and his own jungle and cabin....*
>
> (Burroughs 1983: 244–49)

The "exultant freedom" Tarzan feels on leaving the "narrow and circumscribed sphere" of civilization whose "restrictions and conventionalities" hem him in (Burroughs 1983: 249) raises a number of interesting questions with direct relevance to this volume. Is Tarzan having an "adventure"? What is the relationship between adventure and civilization? Does Tarzan's experience reflect a broadly held late nineteenth and early twentieth century cultural attitude that modern life was constraining and unsatisfying? Was adventure, in its many guises, seen as an antidote to this malaise?

To address these questions this paper examines Edgar Rice Burroughs's *Tarzan of the Apes* (1912) and Georg Simmel's essay "The Adventure" (1911), considering them in the context of arguments made by T.J. Jackson Lears (1981, 1983, 1984). The paper argues that late nineteenth and early twentieth century upper-middle-class interest in adventure was part of a zeitgeist reflecting a desire to escape an urban, modern, industrialized life that was perceived as stale, mundane and, in a way, unreal. Concern with adventure was a response to and a manifestation of this "crisis of modernity." The popular success of Burroughs's Tarzan is evidence that the working classes shared these concerns.

Inverting Geertz's assertion that "Doing ethnography is like trying to read (in the sense of 'construct a reading of') a manuscript—foreign, faded, full of ellipses, incoherencies...." (Geertz 1973: 10), I read text as culture, examining "adventure" in different genres—adventure stories, social theory, and serious literature—within a context of historical and cultural commentary.

Along with Burroughs's *Tarzan* and Simmel's essay, I draw on two works from that period's canon, *Civilization and Its Discontents* (Freud 1930) and *Heart of Darkness* (Conrad 1899) and two less-known texts, written by men who lived adventurous lives, Erskine Childer's *The Riddle of the Sands* (1903) and Sir Arnold Wilson's *Southwest Persia: Letters and Diary of a Young Political Officer* (1940). These latter texts provide an alternative, parallel treatment of adventure that de-emphasizes the importance of the individual, emphasizes the importance of service in a larger cause, and downplays danger and extreme experience even as it describes it. The contrasts further our understanding of adventure and perhaps of anthropology's place in it.

Adventure in the Zeitgeist: Tarzan and Simmel

Simmel argued that "the most general form of adventure is its dropping out of the continuity of life ... an adventure stands in contrast to that interlocking of life-links, to that feeling that those countercurrents, turnings and knots still, after all, spin forth a continuous thread ... it is a foreign body in our existence which is yet somehow connected with the center..." (Simmel 1959: 243). However, for Tarzan hunting is hardly "out of the continuity of life"; "[H]e had done it a hundred times in the past" and, moreover he has only undertaken to do it once again because he has "a reason" (10,000 francs) that has replaced the usual one, hunger. So one wonders, is Tarzan's hunt an adventure? Simmel's assertion, that "The adventure is freed of the entanglements and concatenations [of everyday experience] … and is given a meaning in and of itself" (Simmel 1959: 244), makes it even harder to see Tarzan's hunt as an adventure—unless we also wish to see our own repeated trips to the supermarket as adventures.

Simmel also says, "… in the adventure we abandon ourselves to the world with fewer defenses and reserves than in any other relation…" (Simmel 1959: 248). Burroughs uses the responses of Tarzan's fellow Europeans—his challengers—to show how extraordinary Tarzan's actions are. They cannot imagine entering the jungle unarmed; their idea of adventure is to go "out with a couple of rifles and a gun bearer, and twenty or thirty beaters" for their "increased safety." They will break the continuity of their existence, but they will contain the risks and limit the degree to which they face the world with "fewer defenses and reserves." Tarzan deserves our esteem because he does what we (and his colleagues) cannot and also, I suspect, because, though he is more confident of his skills, he also risks more. When we imagine him naked and armed only with his rope and knife, Tarzan's hunt does seem more like an adventure. Still, what would be an adventure for us is more ordinary for Tarzan.

We are more certain that Tarzan is having "an adventure" when we read Simmel's assertion that, "the adventure is again and again not the substance which it offers us and which, if it were offered in another form, perhaps would receive little heed, but rather the adventurous form of experiencing it, the intensity and excitement with which it lets us feel life in just this instance" (Simmel 1959: 253). It is precisely this "intensity and excitement" of "exultant freedom" that Tarzan experiences when he enters the jungle.

In sum, Tarzan engages in activities that would indeed be adventures for us but are for him almost mundane. His extraordinary abilities and feats certainly help explain the popularity of his story. But may one still not ask why the "narrow and circumscribed sphere" of "civilization," of which clothing was but one element, literally so hemmed him in that he could not feel alive, and why Tarzan had to return to the jungle to "feel life"? One answer is that both Simmel's notion of adventure and Burroughs's Tarzan fantasy reflect the spirit of their age.

Tarzan exemplifies arguments Freud makes in *Civilization and Its Discontents*. Back in his trees, Tarzan realizes that in civilization he had been "hemmed in by restrictions and conventionalities." This realization will ultimately throw him into emotional conflict, but his first feeling is "exultant freedom." Had he read Freud, Tarzan would have known that his feeling echoed Freud's "contention … that what we call our civilization is largely responsible for our misery, and that we should be much happier if we gave it up and returned to primitive conditions" (Freud 1961: 33).

Freud argued that "civilization is built upon the renunciation of instinct" such that it "presupposes precisely the non-satisfaction (by suppression, repression or some other means?) of powerful instincts," and he argued that the result is that civilized man feels "frustrated" (Freud 1961: 44). Indeed, "a person becomes neurotic because he cannot tolerate the amount of frustration which society imposes on him in service of its cultural ideals," and Freud adds, "… it was inferred from this that the abolition or reduction of those demands would result in a return to the possibilities of happiness" (Freud 1961: 34).

Both Simmel and Freud stressed the human desire for the new and different, Freud suggesting that "happiness derived from the satisfaction of a wild instinctual impulse untamed by the ego is incomparably more intense than that derived from sating an instinct that has been tamed" (Freud 1961: 26), Simmel that in an adventure, "continuity with life [having been] … disregarded on principle … we know from the beginning that we have to do with something alien, untouchable, out of the ordinary." (Simmel 1959: 244)

Joseph Conrad too suggested that daily life in "civilization" was something people long to resist or escape. This longing is palpable in *Heart of Darkness* when Marlow (Conrad's protagonist and alter ego) describes how in his youth, seeing blank spaces on a map he would "put my finger on it and say: When I grow up I will go there … a blank space of delightful mystery— a white patch for a boy to dream gloriously over" (Conrad 1988: 11–12). *Heart of Darkness* illustrates the power of the drive for escape, adventure, and freedom; it takes Marlow (and Conrad) into the "Heart of Darkness," and it takes Kurtz over the edge to where freedom and "horror" are conjoined.

The enormous popular success of Burroughs's Tarzan, whose "adventures" clearly spoke to his audience, points to the power of this desire. But why did that popular audience, and Simmel and Freud and so many others in that era feel such a need for something more than everyday, modern life? The answer must lie in the character of that modern life and the way it was perceived by those who lived it.

Freud, writing not so much as an analyst of others but as a subtle observer of his own concern argues that the happiness brought by technological progress is illusory. Others see an even darker prospect. Where Freud is disenchanted with the notion of progress, others see industrialization, mechanical reproduction, photography and lithography and advertising marking the loss of "authenticity" so that "A dread fascination with deceit and imposture pervaded nineteenth century American culture" (Lears 1981: 35 and see Orvell 1989; Bendix 1997).

This contrast of the authentic with the inauthentic is a key theme in both Tarzan and *Heart of Darkness*. Through the repeated contrasts between Tarzan—the real Lord Greystoke—and Clayton—the (unintentional) usurper of his title, Burroughs highlights how "conventionalities" are necessary in society but not really expressive of what one is or wishes to do. For example, we read that "Tarzan would not ruin good food in any such foolish manner [that is cooking it], so he gobbled down a great quantity of the raw flesh.... And then ... wiped his greasy fingers on his naked thighs and took up the trail...." while "in far-off London another Lord Greystoke, the younger brother of the real Lord Greystoke's father, sent back his chops to the club's chef because they were underdone, and when he had finished his repast, he dipped his finger-ends into a silver bowl of scented water and dried them on a piece of damask." (Burroughs 1983: 78). Burroughs's sympathies lie with Tarzan, and no one has less "artifice" or lives a purer, less affected state of being than he. When Tarzan eats, it is really *au naturel*.

Conrad too draws attention to the authentic and the bogus. Marlow says of the European outposts he passes sailing down the west coast of Africa,

> Every day ... we passed various places—trading places—with names like Gran' Bassam and Little Popo, names that seemed to belong to some sordid farce acted in front of a sinister back cloth ... the uniform somberness of the coast seemed to keep me away from the truth of things within the toil of a mournful and senseless delusion.... Now and then a boat from the shore gave one a momentary contact with reality. It was paddled by black fellows ... they had bone, muscle, a wild vitality ... they were a great comfort to look at. For a time I would feel I belonged to a world of straightforward facts; but the feeling would not last long (Conrad 1988: 16–17).

The Fin de Siecle Zeitgeist of Malaise

All men may have felt diminished by modern life; however, their experience of loss had class-specific characteristics. If Tarzan's fierce defense of what he valued, his physical prowess, and his vigor spoke to the psychic needs of the emerging managerial professional class, the broader, popular attraction of Tarzan must also be recognized and explained.

Marx (1978[1844]) argued that industrial capitalism led to the alienation of the working class, and within the vast literature dealing with this issue one finds the argument that the rise of manufacturing, mass production, and mass consumption created a challenge to male identity, as the question of how to *"continue the manliness hitherto bred in war and work"* became *"a central cultural problem"* (Livingston 1998: 423, emphasis added).

Lukács, Simmel's student (1971: 88, 97), proposed a workingman emasculated by proletarianization, experienced in part as rationalization and bureaucratization of labor process and a consequent loss of control over time, labor, and product. Having lost control of the work process, the former artisan or small producer has also lost "the site of self-discovery" as "genuine selfhood can no more be derived from the abstract, unskilled social labor of the fully mechanized workplace than it can from the sluggish daily routine of the private household" (Livingston 1998: 418). In short, the mechanized laborer, his time and his labor no longer his to regulate, has lost his autonomy. Subjected to someone else's control, he had become feminized. Workers at the New England Bolt Company described their encounter with scientific management in 1913 in precisely these terms, "If the Taylorisers only had an apparatus that could tell what the mind of the worker was thinking, they would probably develop a greater 'efficiency' by making them 'cut out' all thoughts of being men" (Livingston 1998: 421 quoting D. Montgomery, *The Fall of the House of Labor* 1987: 220–21).

If workers were alienated and as a result felt emasculated, if "modern industrial routine numbed workers' minds and drove them to waste their energies in 'sedative pleasures' that were 'irrational and extravagant,'" then a key question was how society could "utilize the workingman's latent vitality in order to increase his industrial efficiency and give to him the rewards of energies, now ineffective, within his body and soul" (Lears 1994: 115)? In practice, the solution was difficult to achieve. However, stultified existence made the quest for, or appreciation of, adventure broadly attractive in the early twentieth century. Vicarious or contained adventure was an anodyne to alienated existence. The popularity of books like *Tarzan*, the rise of amusement parks like Coney Island, and—I would argue—the still-expanding popularity of extreme sports and foods are all evidence of the need for this anodyne.

Members of the emerging managerial professional class experienced different kinds of losses, but for them too modernity meant a loss of vigor, community, and authenticity and an absence of groundedness. Adventure was "real" or at least brought one into contact with the "real," and it is worth situating our understanding of it within the enormously complex and subtle arguments made in a series of major works by Jackson Lears (1981, 1993, 1994). The central frame of his argument is that,

> In the 1880s and 1890s the leaders of the WASP bourgeoisie confronted labor struggle, financial uncertainty, and the even more insidious threat of severe self doubt. They felt cramped, "over-civilized," cut off from real life—threatened from without by an ungrateful working class and from within by their own sense of physical atrophy and spiritual decay (Fox and Lears 1993: xi).

These feelings of unreality came from many sources, including "the corrosive impact of the market on familiar values" (Fox and Lears 1983: xiii), from technological change that "isolated the urban bourgeoisie from the hardness of life on the land," from "a softening Protestant theology [that] undermined commitments and blurred ethical distinctions...." (Lears 1983: 10), and from an expansion of the market that left "more and more people ... enmeshed in the market's web of interdependence," making "liberal ideals of autonomous selfhood ... ever more difficult to sustain." (Lears 1983: 7)

Some solutions compounded the problems. Lears notes the rise of "nostalgia for a pristine 'natural' state" (Lears 1983: 23) that was, ironically, often provided in a "commodified version of the past served up in national advertisements" (Lears 1983: 23). He argues that consumption became a vehicle to replace lost autonomy (Lears 1983: 29) though people were not satisfied by these substitutions (Walter Lippman, for example, saw feminization arising as the economy shifted from being focused on production—a masculine activity—to consumption—a feminine one [Livingston 1998: 420ff.]). The growing commercialism and expansion of media exacerbated the very feelings of unreality whose cures they touted (Lears 1983: 21). As a result, "never before had so many people felt that reality was throbbing with vitality, pulsating with unspeakable excitement, and always just out of reach … the feeling of unreality helped to generate longings for bodily vigor, emotional intensity, and a revitalized sense of selfhood" (Lears 1983: 10). Thus at the turn of the last century many people felt they were living in "A weightless culture of material comfort and spiritual blandness … breeding weightless persons who longed for intense experience to give some definition, some distinct outline and substance to their vaporous lives" (Lears 1981: 32). Works like Tarzan provided a vehicle for imagining an alternative—it is hard to imagine a more complete rejection of "urban luxury" or a more complete abandonment of civilization's amenities (recall Tarzan's horror at cooked food) than Tarzan's life among the apes.

Renewed militarism was also a central theme in turn-of-the-century antimodernism. Its hallmarks were seeing positive or constructive characteristics in violence, ferocity, passion, and strenuous action, of which war was the archetype. "Anti-modern militarists … equated the decline of ferocity with encroaching enervation" (Lears 1981: 101). They believed that the desire for peaceful commerce reflected a loss rather than a gain. It was in this context that William James correlated the impending demise of the "manly virtues" with "pacific cosmopolitan industrialism," a stage of development in which an older "pain economy" was giving way to a "pleasure economy," described as "a world of clerks and teachers, of coeducation and zoophily, of 'consumer's leagues' and 'associated charities,' of industrialism unlimited and feminism unabashed" (Livingston 1998: 423).

In this bland world, "life at war … seemed to promise authentic experiences no longer available in everyday life: the opportunity for physical and moral testing, the sheer excitement of life amid danger and death … the chance to escape the demands of bourgeois domesticity and reintegrate a fragmented sense of self by embracing a satisfying social role" (Lears 1981: 98). By the 1890s "the martial ideal emerged as a popular antidote to over civilization … it animated cults of strenuosity and military prowess…" (Lears 1981: 100). In this revised view of warfare and conflict, men applauded "spartan virtues," exemplified by "Theodore Roosevelt's cult of the strenuous life," and "the risk-taking demanded by the strenuous life … became a path to psychic well being" (Lears 1981: 102). Tarzan speaks to this as well. Repeatedly toward the end of the book as Tarzan struggles to win and protect Jane, his "primal forces" surge to the fore. For instance, as he strives to free Jane from an entanglement with Canler (who is the very embodiment of the market forces and market morality that antimodernism questioned), Jane tells Tarzan, "This is not an African jungle"… "You are no longer a savage beast...." To which Tarzan replies, "I am still a wild beast at heart." It is no surprise then, that when Tarzan and Canler meet and Canler makes the mistake of grabbing Jane, Tarzan's primal instincts burst through and Canler finds, "a heavy hand closed upon his arm with a grip of steel ... and in a moment he was being shaken high above the floor, as a cat might shake a mouse" (Burroughs 1983: 269–70).

Tarzan's ferocity, independence, and competence at violence power-fully reflect the antimodern values; even the exploits of Teddy Roosevelt—the exemplar of the strenuous life, who hunted game like the Europeans who wager with Tarzan—pale beside Tarzan's. No one is less cut off from nature than Tarzan, "A personification was Tarzan of the Apes, of the prim-itive man, the hunter, the warrior. With the noble poise of his handsome head upon those broad shoulders, and the fire of life and intelligence in those fine, clear eyes, he might readily have typified some demigod of a wild and warlike bygone people of his ancient forest" (Burroughs 1983: 97). As a near god, a paragon of physical perfection raised and living in conditions of unimaginable rigor and action, Tarzan is the very antimatter of "physical atrophy." If modern life made men physically and morally soft, circumscribed and interdependent (Lears 1983: 10), then Tarzan was the imaginary antidote. Totally embedded in a landscape considered far more dangerous and savage than that of the United States, totally autonomous and independent, totally committed to an (implicit) Darwinian ethos that was absolute in its all-encompassing reality, Tarzan exemplified a life that was everything that modern middle-class life was not. And from our con-temporary perspective, no one who has killed hundreds of lions can be ac-cused of "zoophily," unless it is as a matter of taste. Simmel's essay on adventure in (a settled) modern life may lack the drama of *Tarzan*, but it shared the same impetus.

Adventures in Reality: Childers and Wilson

In *Tarzan*, Simmel, and Freud, adventure is characterized by a longing to escape ordinary life defined as dull, mundane, and stale. But how did men living extraordinary lives see and describe adventure? Were they shaped by the same *zeitgeist* or a different one? Arnold T. Wilson, whose letters and diary date from 1907 to 1914, and Erskine Childers, who published *The Rid-dle of the Sands* (1903), were near-contemporaries of Burroughs and Simmel, but their response to "adventure" seems less excited, flatter, and more controlled than that of their more impassioned contemporaries.

In the early 1900s, Wilson was a junior political officer in Iran, doing extensive reconnaissance work among nomadic pastoral tribes during very unsettled times; he died in 1940—at the age of 56—serving as a gunner on an RAF bomber (Marlowe 1967). Childers fought in the Boer War, was later drawn to the Irish Republican cause and smuggled arms into Ireland; he nonetheless volunteered for the Royal Navy, became a pilot and won a DSC. During "The Troubles" he was taken and executed by an Irish Free State firing squad on 24 November 1922 (Boyle 1977).

Are their extraordinary, adventuresome lives reflected in their accounts of adventure, fictional or otherwise?

Ordinary life in *The Riddle* sounds very like the malaise of modernity. The story begins with the young fashion-conscious narrator, Carruthers, at loose ends, feeling "martyred" because he must miss the summer season by staying in London to work. Carruthers's entire life is centered about trivial pleasures; his work at the Foreign Office certainly does not satisfy.

Fortunately, Carruthers is saved from boredom and from himself by a letter from an old school chum, Davies, inviting him to sail in the Baltic. Though the yacht he finds does not meet his expectations and the yachting is unlike anything he has encountered, Carruthers discovers that Davies's yachting has another end, collecting information about the German coast. That, however, does not launch a conventional spy thriller, but something quite different. "And, good Heavens!" (Davies leant back and laughed joyously) "do I look like that kind of spy?" That is, "one of those romantic gentlemen that one reads of in sixpenny magazines, with a Kodak in his tie pin, a sketch-book in the lining of his coat, and a selection of disguises in his hand luggage" (Childers 1995: 75–76). Still, Carruthers knows that "Romance, veiling her face" had handed me the cup of sparkling wine and bade me drink ... instilling ... the gay pursuit of the perilous quest" (Childers 1995: 81–82). Strenuous life again seems the cure.

But the book takes a significant turn. For nearly the next two hundred pages Davies and Carruthers explore the islands and channels of German *Friesland*. No guns appear, no one is shot, there are virtually no chases, even most of the "bad guys" appear to be decent rather than deadly, and the worst of them is more corrupt than evil. Carruthers and Davies sail, go aground, walk the sands, and measure water levels and channels in an area of vast tidal flats devoid of singular natural features, dangerous natives, or other novelistic attractions. We mostly read close descriptions of the region, or at least its riparian portions, and the work involved in charting it.

Davies and Carruthers's fictional work strikingly resembles the "real" work Wilson writes of in his letters and diary. Early on, describing Persia and its attractions, Wilson writes that, "It is, to me, a new and fascinating world.... It is the oldest historical centre on the face of the earth. I am lucky in another way: I know how to survey and there is a vast area here wholly unsurveyed" (Wilson 1940: 22). Wilson constantly repeats this theme. Much of his time was spent "either working at reading and writing Persian, or talking the local dialect, or in surveying and compiling detailed reports for military purposes" (Wilson 1940: 29–30). It is what Scott calls "seeing like a state" (Scott 1998). The work is, allowing for the difference between the sea and the desert, remarkably like that of Childers's heroes. For Wilson, the attractions are physical exertion (the transatlantic version of Roosevelt's cult of the strenuous life), competence, and seeing something "new" (Wilson 1940: 45–46). And the outcome is "material for the Intelligence Branch" (Wilson 1940: 56). For, in fact, it turns out that gathering information is the task at hand; it was seen as what had to be done, was to be done, and should be done, "... nothing carelessly ... never in a hurry; if need be ... 16 hours on end, doing thoroughly what ... might have done in an hour superficially. Every local notable ... indexed and cross-indexed; every town and village ever mentioned in current dispatches and telegrams identified and placed on a map ... every old file searched..." (Wilson 1940: 121).

Now, one may wonder why Wilson was not bored. First, and probably most importantly, he (and one presumes others like him) took the gathering of information seriously and wanted to do it, "I think I earn my pay; the work has a spice of danger but it is of absorbing interest. Maps and reports occupy every hour I can snatch from endless conversation with my hosts. I am proud of the map which covers much territory hitherto marked 'unexplored'" (Wilson 1940: 163).

The "spice" of danger is frequently there: "My tent having several times been pierced by bullets and twice entered by thieves at night, I ceased to use a camp-bed and slept always on the ground; whence it was easy to rise, and easier to hear any suspicious sound" (Wilson 1940: 25). But Wilson constantly "flattens" the adventure through understatement. His descriptions of danger do not dwell on it. Being attacked is "an unpleasant encounter." Should he meet again those who attacked him he thinks with "intent to kill," he will turn his back on them or not speak to them. And, when he must recount these events publicly, he takes the same tack. One finds the same kind of flatness in Childers's account of a Boer War skirmish. He summarizes the outcome of an encounter in which a major, a lieutenant, and several gunners in his unit were killed as follows: "terrible mischief was being done on the ridge..." (quoted from Boyle 1977: 117).

This was clearly the appropriate style, and to some degree the following quote from Wilson lets us know why. "The report I gave of my own treatment outside Shiraz made light of the incident, because I did not wish to get the reputation of being, like one of Napolean's generals, 'a man on whose head tiles are apt to fall'—in other words, someone who is always getting into trouble" (Wilson 1940: 192). Childers's biographer writes approvingly of Childers's history of the deadly skirmish described above, "Remembering that Childers and Williams had faced death together in this small, confused, and isolated action, the dispassionate clarity of their retrospective account was commendable" (Boyle 1977: 117).

Just as Childers gives us a dark view of Foreign Office work in London, when Wilson writes home after receiving orders to return to India that he is likely to become "an Assistant District Commissioner, at the bottom of the civil hierarchy, followed perhaps by a spell as the Third Assistant to a First Class Resident, wearing fine raiment, arranging dinner parties and writing out menus for her ladyship...." (Wilson 1940: 185), his lack of enthusiasm is palpable.

It is of course the danger that made Wilson's (and Childers's) lives different from those of Simmel and Burroughs. And they knew that they were having unique and dangerous experiences. But two things catch our interest: first, that they saw data collection as central to their lives; second, though Wilson and Childers are fully aware of the difference between what they do and a more ordinary existence, instead of glory, what they stress over and again is duty. Wilson stresses this at the outset when he quotes with approval his Colonel's advice, "he said to me, 'Never ask for a particular job, and never refuse one if offered unless it is a soft one'" (Wilson 1940: 18). And he continues in this vein when he writes, "I am just a small cog on a great wheel in a big machine, content if I wear well, do not creak, or pick up grit" (Wilson 1940: 118). He fulsomely describes the works of others in the same vein "I went to see the oil wells ... run by a single British engineer, ... in a howling wilderness, but everything spick and span and showing the pride of a man in his work.... Such men are real missionaries" (Wilson 1940: 282).

But most of all Wilson sees the work as part of a larger project that gives the whole meaning. He establishes this in his 1940 preface to his old notes and letters when he writes, "Before the Great War my generation served men who believed in the righteousness of the vocation to which they were called, and we shared their belief ... we toiled at our own ponderous Gazetteers like willing slaves making bricks for builders yet to come ... we knew ... that though for all of us ... there is one end, yet would our works live after us, and by their fruits we should be judged in days to come" (Wilson 1940: viii). Wilson repeats this theme in numerous passages and it is powerfully echoed when Childers, describing Davies, writes, "the key to his own character ... was devotion to the sea, wedded to a fire of pent-up patriotism, struggling incessantly for an outlet in strenuous physical expression; a humility, born of acute sensitiveness to his own limitations, only adding fuel to the flame." And Davies says of his life, "it's all been wasted till this chance came. I'm afraid you'll not understand how I feel about it; *but for once, in a way, I see a chance of being useful*" (Childers 1995: 88–89, emphasis added).

All this talk of being useful and of endless work in the minutiae of gathering information seems to occur in a place far from ennui, but it takes us back to Simmel, who wrote,

> Something becomes an adventure only by virtue of two conditions: that it itself is a specific organization of some significant meaning with a beginning and an end; and that, despite its accidental nature, its extraterritoriality with respect to the continuity of life, it nevertheless connects with the character and identity of the bearer of that life.... (1959: 246)

Simmel saw adventure as a happy intrusion of difference into an otherwise humdrum life. For Simmel, it is extraterritorial. For Wilson and Childers, life, meaning, work, and adventure are all totally intertwined, to the point that the doings that we think of as adventure, the part that really is exciting and dangerous, is downplayed. It is not surprising then that Childers saw the virtue of his war in having "reduced living to its simplest terms ... realized how little one really needs ... learned the discipline, self-restraint, endurance and patience that soldiering demands ... steeped oneself in the region of fact where history is made and Empire is molded" (Boyle 1977: 97).

Put differently, the men who lived adventurously thought their lives were meaningful because they played a small part in a large, important project. The men who lived more quietly thought adventure came from some grand rupture with the ordinary. Simmel, Wilson, and Childers all saw adventure connected to "necessity and meaning"; however, Wilson and Childers saw it, and indeed lived it, not as external to regular life but as their way of life. Their view confirms Lears's contention and helps clarify the sources and the power of the more deskbound authors' grandiose adventures.

Coda: And Anthropology?

In *Works and Lives* (1988) Geertz somewhat mockingly refers to Evans-Pritchard's flat line account of his WWII East African experiences as "Akobo realism" (Geertz 1988: 61) and provides samples of Evans-Pritchard's work that sound very like Wilson and Childers. Geertz notes that E-P's "tone" was shared by a host of notable British anthropologists (and by Americans too). Meyer Fortes is one British anthropologist mentioned by name. In his autobiographical essay in the *Annual Review of Anthropology*, Fortes describes what must have been a miserable and terrifying experience with a flatness that echoes Wilson and Childers: "It was the policy of the International African Institute that their research fellows should receive some training in the indigenous languages of the areas in which they would be working. As there was in London no one acquainted with the languages of the Gold Coast, I was sent to Berlin, where I worked with Professor Diedrich Westermann for about a month in April and May 1933. It turned out, however, to be a frustrating and indeed rather frightening experience since this was the month when the Nazis were seizing full control of Germany" (Fortes 1978: 7).

Overall, in a manner that resembles Wilson's retrospect on his career, Fortes repeatedly describes himself as a "journeyman" anthropologist, aiming to turn out "a particular product … using the best tools at his disposal" (Fortes 1978: 1). He claims the essay sets out "a few of the more salient developments in social anthropology during the last 40 years," adding that they are ones in which he has been "fortunate enough to play some part" (Fortes 1978: 24). That Fortes (and quite likely Evans-Pritchard and others) shared the imperial attitudes of at least rhetorically downplaying grand adventure and accentuating the importance of submerging oneself in a larger project is not surprising. They were of the British Empire, and extended fieldwork made their lives hardly ordinary.

What of contemporary anthropology? I have argued that both anthropological fieldworkers and colonial officers like Wilson were "instrumental travelers," set off from more casual tourists by the purposive intent of their travels (Bradburd 2000a). This brief discussion of Wilson and Childers reveals another important similarity: the excitement of collecting data, of gathering information, of learning something new and understanding it (not to mention the generic flatness of our own accounts). Speaking personally, when I was working with nomadic pastoralists in Iran, there were moments of fieldwork that were tedium; in others the physical setting and the encounter with something new and different were so exciting I knew I was having an "adventure" in the fullest of Simmel's senses. There were also times where learning something, finding the piece of an anthropological puzzle, or knowing that one had encountered something that would add to the literature was absolutely exhilarating. This leads me to believe that anthropological fieldwork may embody both kinds of adventure: the opportunity for the deskbound academic to move beyond the confines of everyday experience (which it shares with every other mode of seeking and gaining extraordinary experience) and the opportunity to be part of a larger collective project based on painstaking fieldwork and data analysis that is central to anthropology (Bradburd 1998). The first point seems trivially true; I hope I am right about the second.

CHAPTER 4

Tarzan and the Lost Races: Anthropology and Early Science Fiction

Alan Barnard

In the realm of popular literature, the incipient science fiction of the late nineteenth and early twentieth centuries (roughly from 1870 to 1939) is of special interest for anthropology. Unlike the science fiction of more recent times, such early works had terrestrial rather than extraterrestrial settings.[1] Three dominant motifs were those of "lost races," "future wars," and "early man," and some writers, notably Edgar Rice Burroughs, combined all three. These motifs have great potential for the revelation of popular attitudes toward the "other" during the period, and indeed since then. Much of the imagery of such literature constitutes a transformation of the "noble savage" idea, seen through the eyes of writers and readers who were conscious of the implications of evolutionary theory. And indeed, anthropological treatments of alien cultures can themselves be seen as analogues of the "lost race" romance.

Philosophy, literature, and anthropology are conventionally treated as separate forms of writing with independent histories. While I do not deny that they are logically distinct, I have tried to emphasize the similarities, analogies, and transformations that link them, rather than the differences that most certainly exist (cf. Barnard 1989). It would, of course, be stretching the point to regard the eighteenth century simply as an age of philosophy, the nineteenth as an age of literature, and the twentieth as an age of anthropology, but this notion does help us to understand broadly the more complex representations of the "other" prevalent in the popular mind during each respective period.

At a more precise level of analysis, these representations, in fact, can be seen to cross the boundaries of forms of writing in ways that suggest both continuities between them and convergences at particular points in time. This chapter explores the implicit thematic and literary relations between anthropology and the "low" tradition of science fiction, as the noble savage was transformed to "savage other," "lost race," and eventually "self," through the nineteenth and twentieth centuries.

My primary concentration here is on the science fiction rather than the anthropological literature, but parallels will be drawn between specific science fiction works and the broad trends of anthropological thinking and the idea of "anthropology" itself, which emerged and evolved to its present form during the age of early science fiction.

Anthropology and Science Fiction in the Late Nineteenth Century

The term "science fiction" came into general use only in the 1930s. However, modern critics generally agree on an earlier foundation of the genre itself, often with Mary Shelley's (1985 [1818]) *Frankenstein* as the first "true" science fiction (see, e.g., Aldiss 1986: 25–52).

There is no doubt that the late nineteenth century marks the development of many of the themes or motifs that are now regarded as part of science fiction. These were generally known in the late nineteenth century as constituents of the larger genre of "romance" or even (in the case of some works) "scientific romance." Works of scientific romance included those concerning travel to other "worlds" (at first mainly terrestrial, later subterranean and extraterrestrial), "inter-world" warfare, and the social structure of and contact between "primitive" and "advanced" civilizations. Not only were these themes developed in content, but through the vast number and variety of works published during the late nineteenth century, they were also developed in scope. In addition, the mass publication of many such works in relatively cheap editions made possible the spread of science fiction–type ideas far and wide, at least within Europe and North America.

The 1870s mark a period of rapid development for social anthropology too. The small band of scholars who had kept alive the eighteenth-century

principle of monogenesis through the early nineteenth century were vindi-
cated when this idea again became accepted in scientific circles. Bachofen
(1861), Maine (1913 [1861]), and McLennan (1970 [1865]) published their
most significant works in the 1860s, and their impact was soon to reach a
wide portion of the educated public. The year 1871 was a landmark year for
the discipline. That was the date when the mainly polygenist Anthropological Society of London merged with the mainly monogenist Ethnological
Society of London to become the Anthropological Institute, thus marking
both the final recognition of a single origin for all humankind—the principle on which all modern anthropology depends—and the acceptance of
the formerly polygenist word "anthropology," at least in Britain, as a desig-
nation for the general study of humanity (see, e.g., Stocking 1971). The
decade also saw the publication of Darwin's *The Descent of Man* (1871), Tylor's *Primitive Culture* (1871), Lubbock's *The Origin of Civilisation* (1870),
and Morgan's *Systems of Consanguinity and Affinity of the Human Family*
(1871) and *Ancient Society* (1877).

In the late nineteenth century, the emerging discipline of anthropology
had at its core two fundamental scientific bases: the theory of evolution
(both biological and social) and the foundation of ethnographic fact. These
also formed significant elements of early science fiction, which I define as
that form of popular literature that reflects and applies the scientific thinking of the time in relation to the fictional events it describes. Among the
earliest themes in science fiction were the "anthropological" ones of the exploration and recording of the indigenous habitation of the non-Western
world, and, later, of reflections on the nature of early humankind. Strictly
speaking, neither Darwinian theory nor exploration was part of what we
now consider *social* anthropology, but both were important for its eventual
acceptance as a discipline and for its popular understanding. Indeed, 1871
was also the year in which Stanley "found" Livingstone. This, together
with the growing ethnographic knowledge of at least part of the African
continent, helped to form the basis for the wide appeal of H. Rider Haggard's many African adventure stories (e.g., 1885; 1887) and the more fanciful African and South American adventures which were to follow (e.g.,
Burroughs 1963a; Doyle 1936). Within science fiction more narrowly defined, 1871 marks the English-language publication of the first truly "Darwinian" tale, Jules Verne's *A Journey to the Centre of the Earth* (originally
published in French in 1864), a book that was to have much impact on
later writing in both Britain and America.

Three Motifs in Early Science Fiction

For purposes of analysis, I identify three key, early science-fiction themes or motifs that frequently also functioned as subgenres: the "lost race" motif, the "future wars" motif, and the "early man" motif. Other minor themes, less relevant to our present concerns, can be distinguished as well: travel by airship (later, spaceship), discoveries and inventions (e.g., electricity, magnetism, radiation, machine-like humans, and human-like machines), natural and human disasters, and utopian and dystopian societies. Several of these have become prominent in the science fiction of more recent times, and there is obviously considerable overlap between them. The same work of fiction can deal with any number. Indeed the subgenres can overlap considerably, and the most interesting works often have elements of more than one of the categories I define here. Edgar Rice Burroughs's *The Lost Continent* (1963f [serialized as "Beyond Thirty," 1916]) is a good example of "lost race," "future wars," and "early man" rolled into one.

One indicator of the significance of the motif is simply the number of such works included in Clareson's (1984) authoritative, annotated bibliography of 838 science fiction books published in the United States between 1870 and 1939. By my count, some 259 of these can be included under the term "lost race" novels. The majority concern Arctic, Atlantan, and Amerindian societies, though African societies described in the works of mainly British authors publishing in America are included too. Indeed, the paradigm case for me, in view of its immediate impact, longstanding popularity, and special relevance for the image of the white, male adventurer (a forerunner of the paradigmatic anthropologist of later generations), is Rider Haggard's *King Solomon's Mines* (1885). This great classic of imperialist fiction tells of the search for a man lost in pursuit of the diamond mines of the biblical Solomon in the interior of southern Africa.

The next largest category represented in Clareson's (1984) bibliography is the "future wars" motif. I estimate that at least 161 of the works described by Clareson can be included here. Originally, the motif overwhelmingly concerned terrestrial wars, rather than the interplanetary wars of Wells's farsighted imagination or the star wars of more recent science fiction. Most nonracist "future wars" novels before 1940 expressed an apparent distaste for the state of the world as it was prior to or just after the First World War, along with a yearning for a simpler lifestyle. This lifestyle was often a European one transplanted to some distant continent and, very commonly, modeled on some earlier, happier European age. The racist form of the motif, far more blatantly racist than any "lost race" romance, had a different twist. Typically, authors described for their European and American readers scenes of "inferior" Asiatic hordes sweeping across (or tunneling underneath) the Western continents.

The final motif is that of "early man," a category that included contemporary and future, fictional, stone-age societies, as well as prehistoric ones. Among the works listed by Clareson, some 109 may be classified as "early man" stories, including a number which also qualify as "lost race" romances. Among the many writers who dealt in "early man" were Stanley Waterloo, Jack London, and, of course, Burroughs. Burroughs's Carpona and Pellucidar series are primary examples. Pellucidar is an "inner world" of strange animals and barbarian tribes, while Caprona is an Antarctic island where hominid (including human) biological and social evolution can be seen happening before one's eyes. Each world is explored by civilized travelers who, through Burroughs, reveal its secrets to the reading public.

To some extent, the growth in publications in all these motifs during the period is related to the growth of magazine publication and circulation, the serialization of popular fiction, and the consequent release of formerly serialized works as books (see Keating 1991 [1989]: 33–45). "Lost race" works peaked at the turn of the century, at the close of the period of European exploration and active colonization of the African continent, then peaked again in the 1920s, the time of the establishment of social anthropology as a discipline in the British Commonwealth and of the expansion of anthropology in the United States (see Stocking 1989; Kuper 1996 [1973]: 1–65). "Future wars" works peaked in the first decade and a half of the twentieth century, in the volatile period just prior to the First World War. The "early man" motif, thanks largely to the high output of works by Burroughs, made its mark in the 1920s.

Boas began teaching at Columbia in 1896, and Kroeber took the first American PhD in anthropology in 1901 (followed by Lowie in 1908 and Sapir in 1909). Yet it was in the 1920s that Ruth Benedict, Margaret Mead, Robert Redfield, Ralph Linton, and many other famous names of American anthropology began their training in the discipline. In Britain, Malinowski took up his first post at the London School of Economics in 1922 while Radcliffe-Brown inaugurated the study of social anthropology in Cape Town and Sydney in the twenties.

Table 4.1 • "Lost race," "future wars," and "early man: novels from the 1870s to the 1930s, by decade of publication

Date	"Lost race"	"Future wars"	"Early man"	Totals by decade
1870s	5	3	1	9
1880s	21	14	4	39
1890s	53	30	14	97
1900s	54	43	16	113
1910s	21	27	16	64
1920s	74	23	34	131
1930s	31	21	24	76
Totals by motif	259	161	109	

Source: my count and classification, based on Clareson's (1984) summaries of novels in the period

Boas, Malinowski, and Radcliffe-Brown were all public figures, but none as public as Mead. Her first and most important book, *Coming of Age in Samoa* (1928), delved deep into the American subconscious in just the manner of the emerging science fiction of that decade. The subtitle—*A Psychological Study of Primitive Youth for Western Civilization*—could *almost* as easily describe *Tarzan of the Apes*. The original dust jacket illustration of Mead's *Coming of Age* could as easily cover Burroughs's *The Cave Girl* (1963c). In the words of Stocking (1989: 246), it depicts "a bare-breasted maiden rushing an apparently nude young man to a tryst beneath a moonlit palm." Though signed "M.F.," it is remarkably similar in style to the work of Hal Foster, who began as Burroughs's illustrator in the same year (1928), and whose daily Tarzan newspaper strips first appeared a year later.

The "Lost Race" Motif

According to Clareson:

> The basic formula of the "lost race" novel was easily identifiable; an explorer, scientist, or naval lieutenant, either by chance of intentional quest, found a lost colony or a lost homeland of some vanished or little-known civilization (1977: 119–20).

Broadly, the production and popularity of the "lost race" motif followed closely the exploration of the African interior and the development of late nineteenth-century social anthropology. As is usual in this type of fiction, each hero falls in love with the local princess. This success or failure element of plot was to provide the reader with both the required romantic catharsis and the craving to read on, and perhaps read more works by the same author or at least in the same subgenre, as it developed through the last half of the nineteenth century.

The early "lost race" narrative was partly an outgrowth of the much earlier and interrelated genres of "utopian" and "imaginary journey" fiction, but it eventually acquired an independent status and ultimately merged with what we now know as "science fiction." Haggard's *King Solomon's Mines* (1885) became the model for all "lost race" romances, and most of Haggard's later works featured the same characters: Allan Quatermain (the hero, probably modelled on a real-life hunter), Sir Henry Curtis (a dashing young man of Danish, or "white Zulu" blood), Captain Good (a stout, ex-naval man). By their character, all could be anthropologists of their generation or subsequent ones: to me, Quatermain could easily pass for Rivers, Curtis as Malinowski, and Good as Radcliffe-Brown. Within the Quatermain saga as a whole (which includes some fourteen novels and four short stories), ethnographic content is high. Haggard had a knack of being just a step ahead of scientific discovery. As is well known, he described imagined Great Zimbabwe–type ruins before ever having heard of the real thing, just as he anticipated the discovery of gold in the interior by writing about it in fiction. More interestingly, in *Heu-Heu* (1925), Haggard anticipated by a hair's breadth the unearthing of early hominids, most notably type specimen of *Australopithecus africanus* in November 1924.

A number of Burroughs's stories are classics in the "lost race" mold. Although they lack Haggard's fairly high degree of ethnographic competence, they contain all the essentials of Haggard's formula. In the second Tarzan novel:

> Tarzan recalled something that he had read in the library at Paris of a lost race of white men that native legend described as living in the heart of Africa. He wondered if he were not looking upon the ruins of the civilization that this strange people had wrought amid the savage surroundings of their strange and savage home…. "Come!" he said, to his Waziri. "Let us have a look at what lies behind those ruined walls" (Burroughs 1963b: 162–63).

This formula is repeated many times in Burroughs's other Tarzan novels. The City of Opar replaces Rider Haggard's Valley of Kôr. Burroughs's high priestess La takes over from Ayesha, and Tarzan from Leo Vincey, the main protagonists respectively of several Tarzan stories and of Haggard's *She: A History of Adventure* (1927). Burroughs's fictional Waziri (a "good" African tribe who have Tarzan as their chief) replace the Zulu to become the "noble savages" of the early twentieth century. Perhaps the most "anthropological" version of the formula in the Burroughs corpus, however, is not a Tarzan story at all but a Tarzanoid, *The Lad and the Lion* (Burroughs 1963h). The hero, Prince Michael, finds himself on the African shore and makes friends with the locals, both animal and human. Eventually, fed up with European intrigues and in love with an exotic, dark stranger, he renounces the throne that is rightfully his and remains in his now father-in-law's sheikdom, to live as an Arab.

The "lost race" motif is for me the paradigm of both early science fiction and the contemporaneous, emerging social anthropology. It lies at the root of both fictional and factual concerns with the "other" and provides the focus for applying the discoveries of the science of anthropology, or at least ethnography, within early science fiction. The alien "other" has been a necessary narrative device for social anthropology through much of its historical existence, and within science fiction it provides the historical link between the romances of Rider Haggard and the extraterrestrial sagas of more recent writers.

There are even some quite *direct,* if slender, links between anthropology and "lost race" fiction. The Scottish folklorist and anthropologist Andrew Lang was a close friend of Rider Haggard. Haggard dedicated *She* (1927 [1887]) to him, and Haggard and Lang coauthored another fictional work, *The World's Desire,* in which Odysseus is sent on a quest to find the immortal Helen of Troy "in a strange land, among a strange people, in a strife of gods and men" (Haggard and Lang 1972 [1890]: 14). Lang, in fact, was a great influence as a critic and regularly used his column in *Longman's Magazine* to praise novels of romance and adventure at the expense of those of the French realists and their British imitators (Keating 1991 [1989]: 346–47).

In later science fiction, anthropology has left its mark too through an updated, intergalactic "lost race" motif pioneered by Ursula LeGuin, the daughter of A.L. Kroeber. In LeGuin's *The Left Hand of Darkness* (1969), for example, an Earthman ventures to the ice-age planet of Gethen to persuade the inhabitants to join the League of Known Worlds. Like many an anthropologist's notebook, the novel consists of a mixture of ethnographic report, diary, folktales, and factual essays, all between two covers.

In the intervening period, evidence of the link between "lost race" fiction and "lost race" anthropology is conspicuous in Malinowski's famous remark (presumably in reference to style): "Rivers is the Rider Haggard of anthropology; I shall be the Conrad" (quoted in Firth 1957: 6).

The "Future Wars" Motif

At first glance, the "future wars" motif bears much less relation to anthropology than either of the other motifs. Yet there are connections. Time and time again, these fictional conflicts result from racial strife, colonial oppression, and political intrigue. In the nonracist form of the motif, the wise observer, often the narrator, plays a role not unlike the dispassionate ethnographer of the real interwar years. In the racist form, the modern reader will find anthropological wisdom absent, but reflection on its absence can still make some of these texts worth a fleeting glance. Hidden in such texts are ethnographic generalizations, ethnic characterization, and racial stereotypes—above all a distinction between the "self" and the "other" based on a purported attempt to "understand" the latter.

The "future wars" motif, in all its varieties, captured something of the twentieth century that was to come, albeit too often in its most unpleasant aspects. The accuracy of authors' attempts to anticipate in fiction wars that later occurred in fact, and indeed the machinery of warfare, the interethnic strife and political intrigue behind the wars, and the social and political revolutions of the early part of the century, was uncanny. Furthermore, it could be argued that the "future wars" motif represented a kind of *applied social science,* with overtones of Marxism, anti-Marxism, Anarchism, Social Darwinism, and other predictive and prescriptive social theories that followed the rise of historicist approaches in early anthropology. Both evolutionism and diffusionism are implicit in many such works as scientific premises of the causes of the fictional future wars. Evolutionism makes its appearance through the emphasis on "advanced" and "backward" societies, with advancement generally being measured by the twin characterizations of technological sophistication and supposed moral superiority. Several of the novels of George Griffith that glorify the Anglo-Saxon "race" without being overtly racist, fall into this category. Diffusionism rears its head through the threat of the "backward" cultures, often with superior numerical strength, swamping the lands from which the authors and their readers hail. The "backward" nations, though often Asian, can be European as well, and their leaders can sometimes be seen as evolutionary throwbacks or simply as the embodiment of the tacit cultural inferiority of their peoples, thus providing a link even to the "early man" motif.

Paradoxically, primitivism lies implicit in the "future wars" theme too, for there is often a longing for a time when the mad dictators and their dreadful weapons of destruction were not present in the world. The "future wars" motif was important too simply by virtue of the huge number of writers who delved into it. More than one hundred different authors produced terrestrial "future wars" novels in the United States in the four decades before the First World War.

The "Early Man" Motif

The relics of "early man" came to prominence only in the latter half of the nineteenth century. A brief look at the fossil record from the point of view of the history of discovery—as opposed to the place of the fossils in the modern conception of human evolution—is revealing.

"Neanderthal Man" was discovered in Prussia in 1857. In 1886 and subsequently, further Neanderthals were found in other parts of Europe, and these confirmed early opinions that "ape-like men" had once inhabited Europe. The first "Cro-Magnon Man" was discovered in a burial site in France in 1868 and gave prominence to the notion of "early," but nevertheless relatively modern, culture-bearing Europeans. "Galley Hill Man" was discovered in England in 1888 but was later found to be very modern. "Java Man" was discovered in 1891 and described as *Anthropopithecus erectus* in 1892 and renamed *Pithecanthropus erectus* the following year.

After 1900 a number of Neanderthal and Cro-Magnon-type fossils were discovered in Europe, and these were often given names like *Homo heidelbergensis* (found in 1907) and *Homo primigenius* (found in 1908), along with theories to account for their supposed importance in relation to previous finds. "Piltdown Man" was "discovered" in 1912, described as *Eoanthropus dawsoni* in 1913, and, after occupying for many years a singularly prominent place in the fossil record, revealed as a fake (made up from the skull of a human and the jaw of an orangutan) in 1953.

"Oldoway Man" was uncovered in East Africa in 1913. The specimen was later found to be modern, though the name survived as a stone-tool industry and Olduvai Gorge eventually revealed the richest early hominid fossil deposits in the world. The first tooth of "Peking Man" was discovered in 1921, made public as *Sinanthropous pekinensis* in 1926, and described in 1927. Another tooth originally named *Hesperopithecus haroldcooki* was discovered in 1922. "The Taung Child" was discovered in 1924 and described as *Australopithecus africanus* in 1925. A long and public battle was fought over the relative significance of this *South African* fossil in the relation to the *English* Piltdown Man (see, e.g., Reader 1988).

The development of ethnology or cultural anthropology (very loosely, the study of "lost races") paralleled and at times intersected the development of human paleontology (the study of "early man"). Haddon's *History of Anthropology* (1910), still the finest of all monographs on that topic, treats *physical anthropology* (including human paleontology, comparative psychology, and racial classification) along with *cultural anthropology* (including comparative ethnology, sociology, linguistics, environmental geography, and archaeological and ethnographic studies of technology) as two aspects of the same science. Lubbock, in his time, undoubtedly the most prominent of all anthropologists in the public mind, devoted his first major work to "early man" (1865) and his second to "lost races" (1870).[2]

The first recognizable "early man" *novel* was probably Austin Bierbower's *From Monkey to Man* (1894), which tells of the evolution of the "Missing Link" after he was forced by a glacier to abandon his northern paradise. More significant for the development of the motif was Stanley Waterloo's *The Story of Ab* (1897). In this book, Ab discovers the use of fire, invents the bow and arrow, and learns to swim. Encapsulated within this single character is the wondrous advancement of humankind from the Palaeolithic to the Mesolithic stages and beyond. Similarly, Jack London's nameless narrator in *Before Adam* (1906) dreams humankind's past through his latent, racial memory, and thus again embodies the evolution of human society.

Other notable works in the same mold include Wells's "A Story of the Stone Age" (1976 [1899]) and Burroughs's *The Eternal Savage* (1963e). In Burroughs's story, the heroine is transported by earthquake to her ancient love Nu of the Neocene and the chapters alternate between present and past time. More dramatically, Burroughs's most famous character, Tarzan (who in fact has a cameo role in *The Eternal Savage*), took on the role of "early man" personified, in his case, as a modern man metaphorically thrown back to the "age" of the ape. Tarzan (Burroughs 1963a) is both a "missing link" and a European traveler among alien "primeval" hominids and "primitive" humans. He crosses the boundary between "early man" and "lost race" fiction, and his existence tempts the reader to reflect on the anthropological questions heightened by African exploration and the ensuing culture contact.

In light of Tarzan's savage and noble masculinity, it is worthwhile to reflect too on the inner strengths of some of Burroughs's other characters. An obvious comparison is the female "Tarzan," Nadara, who is herself cast in contrast to the bumbling and feeble male, "Jane," figure, Waldo Emerson Smith-Jones—a well-educated man (except in practical matters), well read (except in fiction), and a master of languages (especially dead ones). These characters are the subject of *The Cave Girl* (1963c), whose first part was published in 1913, within a year of the first Tarzan tale. Waldo has long seemed to me about the closest "lost race" fiction has ever got to a *real* anthropologist, though Gordon King, in *Jungle Girl* (Burroughs 1933 [1931/ 1932]), bears strong resemblance to an *idealized* member of our profession. King is a young medical doctor who decides "to devote himself for a number of years to the study of strange maladies," and in the course of this devotion he ends up rescuing and marrying a South East Asian princess and becoming King of Pnom Dhek.

As with the "lost race" motif, the "early man" motif remains in modern science fiction and retains its anthropological touchstone, particularly in its feminine (often marginally feminist) "early woman" form. The best-known example is Jean Auel's Earth's Children series, beginning with the extensively researched (if anthropologically disputatious) novel *The Clan of the Cave Bear* (1980). This tells the saga of an inventive young woman who hunts with bow and arrow in the prehistoric Crimea. There is nothing really new here except the tone. Both the idea of the cave bear itself, and that of a woman who hunts, are found in *The Eternal Savage* too (Burroughs (1963e). Elizabeth Marshall Thomas produced in *Reindeer Moon* (1987) a more anthropologically aware version of the story of "early woman," complete with genealogy and dense ethnographic description. After previous books on her travels in southern and eastern Africa, Thomas successfully transported her own knowledge of the African plains to a fictional ice-age Siberian tundra. A.A. Attanasio (1991) has entered the arena with a story that features a woman with magical powers, localized totemic clans, and ultimately the rise of *Homo sapiens* over the Neanderthals. There are, indeed, a number of other writers waiting in the wings with novels of "early man" and "early woman"—themes (in science fiction and fantasy circles) that now seem set to upstage the last few decades' concern with "sword and sorcery."

Tarzan of the Apes: Noble Savage as "Self"

Consider more closely the example of Tarzan. In *Tarzan of the Apes*, the young boy is marooned in Africa with a tribe of apes. Tarzan's human parents are dead, and he is raised by Kala, a female ape. Unlike real apes, these fictional apes can speak. They also have a considerable degree of *cultural* sophistication—though much less than Tarzan himself later acquires. There is an important contradiction here: *nature is in Tarzan's nurture, and culture is in his genes*. Though raised by wild apes, he was born an English aristocrat and, in a way, inherits his civilization. The imagery of the Tarzan stories is dependent on this reversal of the nature/culture opposition, and ultimately it made Tarzan the literary achievement he was.

The major English critical works on Edgar Rice Burroughs are by a science-fiction writer (Lupoff 1965) and a classicist (Holtsmark 1981, 1986), and both give special consideration to his sources. Lupoff is mainly concerned with sources for the idea and exploits of Tarzan in imperialist fiction, while Holtsmark concludes that both the style and the plots have their origins in ancient Greek and Roman sources. For me, though, the true context of Burroughs's work lies in the literary, political, and academic milieu of the early twentieth century. It lies in Tarzan and other Burroughs characters as reflections of the past and as images of the present, a theme also touched on by some of the many other commentators who have looked at the Tarzan phenomenon: from French literary critic Francis Lacassin (1971), to British social anthropologist Rodney Needham (1983), to Nigerian-American Funmi Arewa in a Berkeley PhD on the topic (1988).

Tarzan was more than a character in juvenile fiction. He rapidly came to symbolize the most desired objects of Western culture, especially American white male culture: power, youthfulness, and rugged individualism. These same objects are engrained too in the image of the anthropologist of Burroughs's time, a figure who perhaps had more in common with Tarzan than with any comparable character in fiction. Unlike most members of his race, Tarzan could roam freely through the African jungle on adventure after adventure. He could also catch a ship or plane to visit London, Paris, or New York, and sometimes did. He was highly regarded by his African friends, whose languages he spoke fluently. It is a peculiar irony that the *Hollywood* Tarzan was reduced to an inarticulate dummy, an image from which he has never recovered (cf. Essoe 1968: 70–73; Porges 1975, vol. 2: 779–83).

In American historiography, there is another important parallel, brought on by Frederick Jackson Turner's early paper "The Significance of the Frontier in American History" (1920 [1893]). Turner argued that the American national character is derived not from European culture but from the American self-identification with the western frontier. Even city dwellers, he suggested, could identify with the denizens of the forests, plains, or mountains at each phase of American history. For our purposes, it does not matter much whether Turner's ideas still hold true, or even whether his specific points really ever held true. Of interest is that the Turner thesis became the accepted theory of the development of American nationhood and that this theory dominated American historiography from the 1890s to the 1930s. Both science fiction and anthropology have elements of this frontier myth at their very core. Science fiction throughout its history has been concerned with defining the commonplace or analyzing the familiar through the unknown. Anthropology, especially in the United States and countries influenced by the American tradition, is premised on a similar need to understand one's own culture by studying other, generally more "primitive" cultures, on the fringe of the social universe of Western society—whether in the "wild west" or on some far away "lost continent."

Science fiction shares with anthropology the character of "adventure" (*Abenteuer*), in Simmel's complex sense of that term. In 1910, as the real "future war" loomed and the real "early man" and "lost race" discoveries stood not far away, Simmel wrote: "An adventure is certainly a part of our existence, directly contiguous with other parts which precede and follow it; at the same time, however, in its deeper meaning, it occurs outside the usual continuity of this life" (Simmel 1997: 222). An adventure thus has elements of both the familiar and the alien, the contiguous and the far away, the "self" and the "other." This is what gives science fiction, in general, its attraction. Certainly, it is one element of the Tarzan phenomenon. Arguably, it is also what gives anthropology its own special attraction, both then, for example in the young Malinowski in his pursuit of Trobriand custom and English style (Kuper 1996: 1–34), and now. Indeed, it gave anthropology an edge on other growing social sciences of that time in the recruitment of some of the most imaginative social scientists, in virtually all then current traditions, including German-Austrian, French, British, and American, among others (e.g., Barnard 2000).

Noting the parallels between forms of writing as distinct as eighteenth-century philosophical anthropology, nineteenth-century travelogue, turn of the century historiography and science fiction, and twentieth-century ethnography can, in fact, provide insight into the "self"/"other" problem. It is precisely anthropologists' semi-detachment from their own cultures that has provided the discipline with its understandings of human nature. At the same time, it is anthropologists' vivid portrayal of the exotic (not infrequently with anthropologist as Tarzan figure) that has given the discipline its vitality in the eyes of outsiders and its practitioners alike. This is true not only of ethnography, but of theoretical works in anthropology as well. In many of Lévi-Strauss's works, for example, an implicit "noble savage" comes to occupy the positions of both the "other" and the "self" (the anthropologist), as the ambiguous title of his most famous work, *La Pensée sauvage* (1962), testifies.

The "noble savage" is a pervasive symbol in anthropological thought. Especially in eighteenth-century Europe, he personified the natural condition of humankind. The presence of a "noble savage" (traditionally male, or at least grammatically masculine) was often more allegorical than ethnographic, and this made his image both flexible and powerful. Such an image was, and is, virtually a necessary fiction for anthropology, for it is through this image that anthropology defines both itself and its object. Many anthropologists today would probably deny its presence in their work simply on the grounds that there never was any such thing as a noble savage, much less a lost race or a Tarzan of the Apes. Yet Tarzan and the noble savage are there all the same. Ethnographic descriptions are built upon an opposition between the culture of the observer and that of the object, often even when the anthropologist works "at home" (Jackson 1986). The discovery of strangeness in the familiar is but a return to the hidden metaphor of the noble savage, or in its nineteenth-century transformation, the lost race. In anthropological theories that differentiate "primitive" from "non-primitive" societies (such as evolutionist ones), the noble savage survives as the representation of virtue in the exotic. In anthropological theories that do not make this distinction (such as relativist or postmodernist ones), he survives as a reflection of the common humanity at the root of all cultures.

Conclusion

Science fiction as a genre paralleled and followed the great advances in science that have occurred in the last century and a quarter. More specifically, the subgenres that emerged as among the most popular that popular literature has ever produced mirrored the rise of anthropology in the early part of the science-fiction era. Debates over the relation between the races, heredity versus environment, the evolution of sociopolitical systems, and the development of religion were all replicated in the popular literature of the late nineteenth and early twentieth centuries (Street 1975).

While the "future wars" motif captured one aspect of the public imagination—the potential mini-age of destruction that was soon to encompass European civilization—the "lost race" and "early man" motifs captured the primitivist urge behind their fears. "Lost races" and "early man" were the images of the new sciences of ethnography and anthropology (including prehistoric archaeology) at their greatest periods of popularity—the late nineteenth and early twentieth centuries. It is perhaps no accident that among the most popular novels today, are those of Jean M. Auel, who stands in relation to feminist, interpretivist, reflexive anthropology (e.g., Okely and Callaway 1992) in much the same way as H. Rider Haggard and Edgar Rice Burroughs did in relation to the anthropologies of their own, colonial, evolutionist, and positivist eras.

Commenting on the relation between the imperialist novels of the Rider Haggard school and the anti-imperialist South African novels and stories of more recent times, Hammond and Jablow suggest:

> In general there is an increased awareness of cultural variation among the Africans manifested in frequent and accurate use of ethnographic detail. In a sense the [fictional] literature reflects the coming of age of anthropology, since the authors pride themselves on the authenticity of their descriptions of tribal customs and history. But modern anthropology only adds detail *to the fundamental conceptions which remain traditional.* (Hammond and Jablow 1970: 120; my emphasis.)

Hammond and Jablow, of course, refer to fundamental conceptions of fiction, but what they say might as easily apply to anthropology itself. We anthropologists have not jettisoned our notions of "primitive society." Rather, we have transformed them to meet the needs of the discipline itself and perhaps too of the societies from which we come (cf. Boon 1982; Kuper 1988).

Indeed, we may not be fully aware of the transformations of the primitive illusion, or of the fact that it has been recycled back to anthropology. It comes back to us through the fiction we now read when not on our professional guard, and even in the fiction we, as children, once read so innocently.

Notes

1. An earlier version of this paper was published in Eduardo Archetti's *Exploring the Written* (Barnard 1994). I am grateful to Universitetsforlaget AS (formerly Scandinavian University Press) for permission to publish this revised version. References to the more arcane literature are to be found in that version.

2. Apart from his amateur career in anthropology, Lubbock was a prominent banker and Liberal Member of Parliament, a Trustee of the British Museum, Treasurer of the Royal Society, and Vice-Chancellor of the University of London. He served as President of the Ethnological Society of London and of the Anthropological Institute. Already a baronet, Lubbock was elevated to the peerage as first Baron Avebury in 1900 and died in the year of the Piltdown "discovery."

CHAPTER 5

Avant-garde or Savant-garde: The Eco-Tourist as Tarzan

A. David Napier

The Call of the Wild

> I don't feel safe in this world no more.
> I don't want to die in a nuclear war.
> I want to sail away to a distant shore,
> And make like an ape man.
>
> The Kinks, "Apeman"

Every adventure novel at some level owes its successes to our shared perceptions about how the dignified person (as far back as Homer's *Odyssey*) behaved under trying circumstances in alien places. By the eighteenth century, such trials were socially formalized in what came to be known as the grand tour, an extended journey in which any gentleman of means was meant to show his mettle by carrying his manners to places where they might be unusually tested. Like a controlled experiment in the laboratory of life, such journeys depended upon specific and sometimes extreme conditions.

In a provocative collection of essays on travel and travel narrative, Paul Fussell reminds us of the connection between our words "travel" and "travail"—that is, the similarity between the excitation involved in the experiencing of something strange or unsettling, and the strain, the emotional agitation, and even the agony that is traditionally a part of being away from home. These intense emotions—which, as Fussell reminds us, Byron once compared to those experienced in gambling and battle—induce sufficient anxiety among those who travel that travelers are bound, by the very nature of the endeavor, to make exaggerated claims about what they have seen and done. If those who travel can resist the temptation to lie outright, at least their successes will depend upon the ability to visit their narratives with sufficient doses of pure fiction. Otherwise, Fussell maintains, no one will believe what they say.

Travel writing requires, in other words, that authors not necessarily lie outright, but at least convince their readers that the experiences that are rendered in narrative are believable and that they are exceptional—that what the author actually did was beyond the ordinary, and that his ability to complete his passage rite makes him, as it were, "experienced." This ability to convince people that one understands their local expectations, but also knows something extraordinary to them, is what Simmel meant by the "unity of nearness and remoteness" (1950: 402) upon which the stranger's powers depend—where "distance means that he, who is close by, is far, and strangeness means that he, who is also far, is actually near" (ibid.). This paradoxical trope, of course, is as much exploited by anthropology's own moral vigilantes as by the merchant classes of European Jews that so interested Simmel. Today, in fact, it is most grotesquely witnessed in the aid industry and in particular in certain forms of medical advocacy where the elaboration of another's travails often includes bombastic self-promotion.

But in nineteenth-century Europe, and especially in the colonial environments where average people could feign nobility, the complex powers of strangeness were more obviously evidenced in how a man of good breeding behaved in extreme circumstances. If our modern-day, rock-climbing (and world-saving) doctor better fits our present "eco-tour" mentality, in former times the taming of the wild by transcending its disordering effects spoke directly to colonial sentiment. Indeed, the rage for formal gardens in England throughout the colonial period attests to how widespread was the desire to tame, and nowhere was the transcendence of the wild better evidenced than in the behavior of a man of good breeding during moments of unexpected duress.

My favorite example of this behavior in *Tarzan of the Apes* occurs even before Tarzan's parents, Lord and Lady Greystoke, ever set foot on "The Dark Continent."

> Short and grisly had been the work of the mutineers [in murdering the captain of the *Fuwalda* and his gang], and through it all John Clayton

[a.k.a. Lord Greystoke], had stood leaning carelessly beside the companionway puffing meditatively upon his pipe as though he had been watching an indifferent cricket match.

As the last officer went down he thought it was time that he return to his wife lest some members of the crew find her alone below.

Though outwardly calm and indifferent, Clayton was inwardly apprehensive and wrought up, for he feared for his wife's safety at the hands of these ignorant, half-brutes into whose hands fate had so remorselessly thrown them.

As he turned to descend the ladder he was surprised to see his wife standing on the steps almost at his side.

"How long have you been here, Alice?"

"Since the beginning," she replied. "How awful, John. Oh, how awful! What can we hope for at the hands of such as those?"

"Breakfast, I hope," he answered, smiling bravely in an attempt to allay her fears.

"At least," he added, "I'm going to ask them. Come with me, Alice. We must not let them think we expect any but courteous treatment." (1983: 12)

Here, the best of us—meaning those properly (i.e., eugenically) bred—are meant to carry our inherited and God-given traits into extreme circumstances where they can be tested thoroughly and, at times, brutally. The sadistic treatment of privately educated boys in Britain's best schools was not only tolerated, but encouraged as a method for conditioning the fiber of young men whose heroism would one day be applauded on the battlefield or in the foreign service. Families like the Huxleys, and even royalty packed their very young off to the likes of Gordonstoun precisely because they could be guaranteed preparation for future trials abroad by lots of ice-cold showers and long marches without overcoats through the worst weather that Scotland might have on offer. In times of peace, it is hard to say what the effects of such socializing processes might be on the likes of Prince Charles; but when times are unstable, those repeatedly tested are meant to excel in ways that prove socially the good effects of Darwinian conditioning.

While Americans seem content to create heroic metaphors from the garment and textile industries ("These colors don't run"), for Victorian and Edwardian British, the chilling of one's blood to a royal blue color provided a better image of how a man of proper breeding ought to cool his emotions when his moral fiber was brought to the test. In fact, a similar argument is to be made when one considers why the French are so obsessed with bringing tamed pets into restaurants and other mannered public places; for showing the world that you can train a dog to behave better at high table than the average member of the working class offers irrefutable evidence of the effects of good breeding in the human kennel. Bringing the exotic home and taming it makes the French bourgeois so much better than that member of the working class, just as allowing a good blue blood to rule over a kingdom of apes makes him so much better than black Africans (who Burroughs does not hesitate to describe as less than ape-like). The language, from the chapter entitled "The Village of Torture," is so vulgar, in fact, that it is embarrassing to read:

> The bestial faces, daubed with color—the huge mouths and flabby hanging lips—the yellow teeth sharp filed—the rolling demon eyes—the shining naked bodies—the cruel spears. Surely no such creatures really existed upon earth.... (1983: 176)

If Burroughs could only have sensed the irony in his own disclaimer.

But irony is hardly a trait of Tarzan. Though what Burroughs is here describing are clearly the racist characterizations of blacks put on by white vaudevillians of his day, the overt intent is to place these tribal peoples beneath the more noble apes now ruled by Tarzan. Indeed, as patriarch and advocate of the natural world, Tarzan becomes perhaps our first popular eco-tourist; for his role is not only heroic but also custodial. Like all of the reactionary politicians these days who have finally agreed to embrace diversity (not on humanitarian grounds, but as a pragmatic means of preserving the gene pool), Tarzan's blue blood entitles him both to be our guide on a fantastic eco-tour, and to be our eco-orchestrator, our deified natural selector who eliminates weaker blacks in favor of preserving the animal kingdom. Thus, while Burroughs's main target was hardly the blue-blooded world he would come both to envy and to imitate at his Tarzana ranch, it took no subtlety for readers to see that Tarzan was much better prepared for dealing with the strange than were either the black Africans he routed, or those blue-blooded Europeans who repeatedly depended upon the wild man's abilities. What a coup, in other words, for the defensive and frightened American reader!

In fact, Burroughs understood very well how to distinguish and critique relative social status through popular epic, and the openly racist nature of the Tarzan legend can be multiply evidenced not only in the manner in which blacks are stereotyped in the many books and films about Tarzan, but by the related racist spectacles with which the story can be connected.

Elmo Lincoln, the "original" Tarzan on film, has a website put up by his daughter on which we learn that Mr. Lincoln also played multiple roles in *The Birth of a Nation,* the historic film that not only, as the website informs us, is "recognized by film scholars as one of the most important films of all time," but is a film widely implicated in the birth of the Klu Klux Klan.

Because travel writing invariably invokes some kind of trial or test—some "travail" that we shoulder as a burden as we journey through life's more challenging moments—travel writing depends on hero epic, either in the form of the person who gets through it all, or in that of the quixotic and antiheroic traveling buffoon. Travel narrative is, in other words, also vulnerable to lampooning. Who remembers Bill Bryson's *Lost Continent* (1989) where the author travels around America in his mother's Chevette, drinking beer in bed and watching TV in cheap motels? What Bryson masterfully captured was the low levels of excitation required by American tourists. On his visit to Mark Twain's boyhood home in Hannibal, Missouri, Bryson alerts us not only to how little it takes to impress an American on the road, but to the absurd delights of complete travel ignorance.

As he looks into the windows, and listens to the recorded messages, at the vinyl and plywood artifice that Hannibal claims as Twain's home, Bryson finds himself following a real American adventure tourist:

> I proceeded from window to window behind a bald fat guy, whose abundant rolls of flesh made him look as if he were wearing an assortment of inner tubes beneath his shirt. "What do you think of it?" I ask him. He fixed me with that instant friendliness Americans freely adopt with strangers. It is their most becoming trait. "Oh, I think it's great. I come here whenever I'm in Hannibal—two, three times a year. Sometimes I go out of my way to come here."
>
> "Really?" I tried not to sound dumbfounded.
>
> "Yeah. I must have been here twenty, thirty times by now. This is a real shrine, you know."
>
> "You think it's well done?"
>
> "Oh, for sure."
>
> "Would you say the house is just like Twain described it in his books?"
>
> "I don't know," the man said thoughtfully. "I've never read one of his books."
>
> (Bryson 1989: 35–36)

In a moment of exquisite irony, the avant-garde musician Henry Kaiser (himself an orphan boy packed off to a military academy by a callous uncle who thought the experience would toughen him) entitled one of his most whacked-out music collections: "Those Who Know History Are Doomed to Repeat It." What this means, of course, is the more you sense what is culturally valued, the more you are compelled to acknowledge the deep power that shared cultural perceptions have over what we acknowledge socially, despite our ability or inability to verify those values personally.

But Bryson's American on tour is no match for anthropology. I remember how surprised I was many years ago when my then thesis advisor, and professor of anthropology, Rodney Needham, informed me that he was prepared to state in *The Times Literary Supplement* (1977: 67) that *Tarzan of the Apes* was one of the most underrated books of all time. Needham had no trouble defending this choice; for in the matter of fiction,

> authors may give themselves away in their figments, or they may reflect the presuppositions or desires of those for whom they write. In either event they often tell us more about conceptions of human nature than would the inevitable qualifications of more didactic ventures (Needham 1977: 21).

Those who write popular fiction, in other words, are going to reflect in a very open manner the concerns and prejudices of their audiences, for otherwise no one would ever read what they write.

Well, so much for badly written ethnography; for doing fieldwork is clearly another matter when it comes to authorial license, since professionalism clearly both defines and limits what readers may be subjected to. Novelists, of course, can be more adventurous. Indeed, in Tarzan's case, there is a good argument to be made for the fact that Burroughs's own professional failures—and especially his experience as a door-to-door salesman—sensitized him not only to the fears of dust-bowl Americans, but also licensed him to display their prejudices in ways that today we find, well, embarrassing. Call it what you will, but march the dusty roads of middle America engaging strangers in polite encounters at the threshold, and you will sense, as the Chevy truck commercial reminds us, "the Heartbeat of America"—that is, the pulse of a nation starving for meaning in a way that no intellectual exercise of Jamesian tone could possibly satisfy. This was, after all, Burroughs's form of "fieldwork." I think here, immediately, of Ed Bruner's tongue-in-cheek study of the tourists he led on a cultural journey through Indonesia. Reminded by Bruner that a princess (with whom the tour group just had a "special" VIP reception) serves such dinners every night, the tourists retort that they don't care about this and ask Bruner to stop ruining their experience (1995).

Mark Twain engaged in much the same kind of "fieldwork"; but Twain was more certain of his intellect than was Burroughs, even if he too had some need to contradict his awareness of the everyday by surrounding himself (in a blatant style that Burroughs would have admired) with the trappings of a bluer bloodline. However, while Twain ridiculed all things overstated (quipping of Wagner that he was "not as bad as he sounds," and at one point of supreme irony, calling him "the Puccini of music"), Burroughs knew that those poor families he encountered in the isolated dust bowls of the American farm belt would be thrilled by, and in fact would demand, to see Rocky-the-Boxer duke it out with the space aliens. Burroughs knew, in other words, just how much the bloated nationalism of our rural poor could be galvanized around a heroic paradigm set to a cartoon of a racist plot.

What is extraordinary about his accomplishment, then, is not its sophistication, but the fact that he had intuitively tapped into a major cultural fear of the foreign and a heroic narrative about how that fear might be conquered through a savage walk in the virgin forest. He did it so well, in fact, that placed side by side with Lord Raglan's magisterial, and much neglected, study of the hero, one readily sees that *Tarzan of the Apes* satisfies nearly every condition set out by Raglan for epic success, including the more obscure understanding—epitomized by, say, Oedipus and Hercules, that a true hero NOT be upstaged by his children. Heroic parents, then, tend to succeed by remaining blind to their children's need for autonomy, while those same children give themselves over to defending environments that their parents seemed (to those children at least) hell-bent on leveling. Over time, the tension gets reduced to cliché or to forms of expression that are culturally policed. In the above-mentioned case, for example, both parents and children express the need to preserve everything wild: when, after all, was the last time you heard someone proclaim to be *against* the environment?

Reduced to being mere props in the epic sagas of their parents, the sons and daughters of the Tarzans, the Clint Eastwoods, and the Ronald Reagans of this world must satisfy themselves as character actors in a laundered version of whatever heroic legend their parents had appropriated for themselves. If as one journalist put it, America is "inside Ronald Reagan," then the hero's descendents must satisfy themselves with the facsimiles of the things that they by definition cannot embody, and, because of this, cannot be. These children become in the end facsimiles of what they have inherited—admirers of a danger now reduced to the safest form of engagement.

So the populace applauds Burroughs's racism in the same way that reality television alludes to the possible truth content of dangerous experiences by orchestrating them artificially in the eco-tour. After all, this is what ethnography does too—at least in the eyes of many indigenous peoples; for there is sometimes little obvious difference between an anthropologist's bunking up with the local clergy, and the average traveler's bunking up at the local tourist trap. Some might feel there's no difference whatsoever, though a number of anthropologists would surely dispute this claim. For many indigenous onlookers, however, the difference between the eco-tourist's safe passage through the now-tamed wilderness and the ethnographer's hanging out with the local gentry in the field is, as we say, academic. What, then, makes the web-surfer's "experience" of the Tarzana, California website any less valid?

By Fussell's and Needham's argument about fiction, it is this last example that is possibly the most honest manifestation of Baudrillard's fictive form of social reality. For Tarzana itself—the Los Angeles suburb that was carved out of the ranch Burroughs bought on the royalties of his writing— builds heavily upon America's real love of *the call of the wild* (as opposed to the uncertainty of the wilderness itself).

Touch the vine on the left of the website's khaki and forest green display and the monkey shakes in turn the files for "special events," "The Tarzana Business Index," and most mysterious of all "the Chamber Gift Store." At the bottom we see Tarzan mounting a lion in an oddly sexual fashion, beneath which we are invited to get more information: "To learn more about this amazing author, be sure to visit www.tarzan.com for some REQUIRED READING." At that site we are treated to a Disney cartoon jungle (including wilderness sounds) atop of which is a display for Disney Online, complete with a Mickey Mouse Disney Movie and Video Finder, and a plethora of facts (running to all of two short paragraphs) that come dangerously close to losing a reader in deep intellectual pedantry:

> As Tarzan® and Jane's first year in the jungle approaches, Jane searches for the perfect gift for Tarzan®. Enlisting the help of her hilarious gorilla and elephant pals, Terk and Tantor, they remind her what an exciting year it's been from outsmarting prowling panthers to surfing lava down an erupting volcano! But that's nothing compared to what Tarzan® has in store for Jane—a surprise that'll show her just how much he understands her world.

Well … what can one say? Given Tarzan's reputation for violence and racism one shudders to think what such a gift might be. In the meantime, we viewers must content ourselves with the creation of our own virtual eco-tour through the darkest rainforests a suburban Californian animator might imagine.

But don't worry; there will be shelter for us when evening falls; for alongside our scholarly story is a DVD bonus feature that allows us to "help build a new tree house for Tarzan® and Jane and then furnish it in [a] two-part game hosted by the Professor." I suppose, like me, you are by now wondering just where we might all go to surrender.

Surrender may actually be a most clever reaction to this pedestrian application of Tarzan's super-humanness; for it is all too clear that Nietzsche was right about the ruthless way we treat an aspiring hero when his or her quest goes awry. One only has to look at the way we license ourselves to scoff at the failures of those who do not fare well in the wild, or who deliberately leave behind the devices that would protect them; for what the eco-tourist retains in terms of safety, the real adventurer throws aside in favor of taking the calculated risk. The former never changes; the latter always does—even if the "change" takes the form of one's own untimely death.

We may easily vent our anger on the adventurer who meets his doom, calling him naïve and unprepared. But we also cheer for the aging eccentric gentleman who decides to row alone across the Atlantic in an open boat, even though we are equally prepared to make fun of his naiveté were any common high sea to have brought his quest to an abrupt ending; for, arguably, the survival of the epic theme means more to us than survival itself—even more than life itself. Asked by a reporter what he thought of J. Michael Fay's recent two thousand-mile trek in sandals across the deepest African jungle (I mean "rainforest"), ninety-one-year-old Sir Wilfred Thesiger answered with his own question:

> "How long did it take him?"
> "Fifteen months…. One difference [the interviewer added] is that Fay had a satellite phone, so if he ran out of food, he could call to arrange a food drop by phone."
> "Well you see," exclaimed Thesiger, "that wreck's it!" (Shnayerson 2002: 106)

One may reasonably ask, what sort of meaning had been wrecked for Thesiger by Fay's retaining his link with civilization? To answer this question it is necessary to understand both why talking about adventure and one's heroic exploits was once considered in poor taste, and why moral exhibitionism has today attracted an entire cadre of anthropologists for whom self-promotion means a great deal.

Small Talk

"Few things are harder to put up with than the annoyance of a good example."

Mark Twain

I can still see him as if it were yesterday. I am fourteen years old and walking with a friend of mine and some other rock climbers through the wilderness in Southwestern Pennsylvania. We are on our way to climb some cliffs in the era when a safety device consisted of a piece of Goldline made into a figure eight that you put your legs through to create a primitive and sometimes dangerous kind of harness. I was too young to be an official member of the Explorer's Club of Pittsburgh, but my explorer uncle (who had just returned from three years alone in the Amazon collecting museum specimens) was an honorary member of this organization and had asked them to allow his somewhat sociopathic nephew to work out his craziness by participating in whatever events they might allow him to.

Because the Club was an ancient institution by American adventure standards (in fact only slightly younger than the New York Explorers Club), and because this was the era before climbing became a popular sport in its own right, being a member of the Club meant that you liked taking all sorts of risks—not only those defined by a personal interest or already-developed technical skill. We all ran rapids, went spelunking, and climbed. And if you were old enough to sign a legal release, you could jump out of planes too. Adventure was a total experience, in other words; and, because there were no satellite photos or global positioning systems, explorers were more concerned about going to places where no one else had been (that year's expedition made the Club the first to visit the Angel Falls plateau) than they were with following climbing routes set out in books with ratings, suggested techniques, and wildlife advisories. On this day, for instance, we would do some climbing, orienteer a bit, and try for the first time to slide down an algae-covered series of rapids and over a waterfall into a pool of ice-cold March spring water.

As we walked toward the cliffs we would climb, the conversation drifted from avoiding the local rattlers to some new things that could be bought from a French-Californian climber who was making hardware on the cheap in his garage. We had blue mimeographed lists of products that he sent in the mail. But that day it just seemed like whatever the rest of the world was doing, it was not rock climbing. There were no climbing stores or books on technique that you could find at your local mountaineering shop because there were no mountaineering shops. There was no money in the sport because there was no market. There was no market because no one had yet invented the kernmantle rope, and climbing (that is to say falling while climbing) was not by and large a very amusing thing. There were also no ready-made harnesses, let alone "bongs" and "friends" and all of the other gadgets that have turned climbing into a safe, highly popular, and profitable enterprise.

I say all of this because I came up quick while walking through the woods that day, raising my eyes to see a man appear almost out of nowhere. He said hello to us and continued on his way. After he left, I was told that his name was Ivan. He was a member of the Club who taught geography at a local high school, and he was out by himself climbing and wandering in the woods for a few days with nothing more than a light blanket and a pocketknife.

I had heard of Ivan already, and a few years later I would see him on the news after he climbed to the top of one of Pittsburgh's many bridges to talk someone out of jumping off. He was on his way home from work when he saw a crowd looking up the suspension cables, so it was nothing for a free climber to walk up the steel incline. He didn't want to say much about that either. And even a few years ago, when I interviewed him for a video project I was doing on climbers of his generation, I asked him about the bridge incident only to find that he was as unwilling to narrate a "nice story for me" about the experience, as he was when the event took place almost forty years ago. Ivan's idea of a challenge was something necessarily done alone, or, if others were necessary, done in the most anonymous way possible. The goal of adventure—and this is the place we connect to the Tarzan story—was to learn something bodily and viscerally, not to talk enough about an "experience" that one might in so doing create a tangible memory.

Sure, there would be a Club presentation after jumping out of an airplane onto the Angel Falls plateau. But talking about what you did personally, like old soldiers after a war, was considered what the British call "bad form." Memorializing what you did through writing about yourself was even more sinful. These were not people given to using allegiance bumper stickers and Purple Heart license plates; because for a certain group of adventurers it would be improper to glorify one's accomplishments—not so much because talking openly might demystify those experiences (though some have made this argument), but because it was believed at a deep level that the meaning was in the doing.

What Ever Happened to Lord Raglan; or, What Is a Hero?

"Seriousness is the only refuge of the shallow."
Oscar Wilde

Having now made my point about the self-indulgent entrapments of adventure narrative, it would, of course, be improper of me to immortalize Ivan or my Uncle Frank by telling you more of what I know of them; but I do want to convey why I have often considered writing about their exploits, only to realize that whatever words I used would be wrong—not wrong because I could never glorify them enough (that is, that their exploits are beyond belief); not wrong because there aren't words to describe their preferred anonymity; not wrong because I might amuse my listeners with their stories; not wrong because describing their exploits would appear chauvinistic; *but wrong* because the kind of embodied meaning they sought out in their experiences was itself of a rather different order than what you or I might imagine while reading a spine-tingling account of a winter on a frozen ice pack or a bivouac at twenty-two thousand feet, or even a thickly described account of another's deep suffering. *Outside Magazine,* in other words, had yet to come indoors, and the idea that everyone would be interested in one's exploits almost diminished their value of a challenging adventure.

So, how do we get from a preference for embodied anonymous meaning in adventure to a preference for narratives about heroic mountain climbers who are followed by hundreds of nameless indigenous peoples to a base camp where our hero-to-be is about to attempt the impossible? Why have these wildly popular narratives replaced, even for many outdoors people themselves, the visceral domain where you are more concerned about what happens experientially than you are about satisfying the photographic needs of the clothing company or geographic society that sponsors your expedition?

Of course, writing about dangerous outdoor exploits is a time-honored tradition that goes back to Homer. But adventure writing as a distinct narrative form only blossoms with the Age of Discovery and the Grand Tour where we witness the sort of writing that would one day culminate in such superhuman achievements as the ascent of Mount Everest by Reinhold Messner—alone and without oxygen. Today, we often see, and even come to expect, that our heroes of the physical use their fame to become political leaders. In Messner's case this meant becoming a wealthy star who now is a leading member of the Green Party. Indeed, we have so much embraced the narrative forms of hero epic that it doesn't even matter if you actually have the experience as long as you can associate yourself with it. As I drove across Interstate 80 this past summer, for instance, I was reminded of this dramatically—one exit for Ronald Reagan's birthplace (the guy who talked about war enough that he actually believed late in life that he had fought in one); another exit for the birthplace of John Wayne. "Come one and all to visit the birthplace of a Hollywood persona." Is it any wonder then that Arnold Schwarzenegger should be elected Governor of California—the super-steroidal bodybuilder-turned-Tarzan-the-Terminator?

But what can it mean to say that these would-be Tarzans are considered heroes? Well, think about the Tarzan myth (the one originally published in *Tarzan of the Apes*) as we review the following list of essential features of heroic narrative described in detail by Lord Raglan in an important book that nobody bothers to read (1979 [1936]). So, here's the list, and if you want to get ahead of me on this one. Just think, as you review this list, of, say, the most wildly popular ethnographies, or, say, the works of our colleagues who consider themselves as "givers of voice" to the weak and needy—who have taken on the duties of monitoring the world's morality, or who feed up to us self-heroizing tropes disguised as Joan-of-Arc guiding those without voice to eternal salvation.

If you wish to follow Edgar Rice Burroughs or Steven Spielberg to the bank, just follow these easy steps:

1) The hero's mother is a royal virgin
2) His father is a king, and
3) Often a near relative of his mother, but
4) The circumstances of his conception are unusual, and

5) He is also reputed to be the son of a god.

6) At birth an attempt is made, usually by his father or his maternal grandfather, to kill him, but

7) He is spirited away, and

8) Reared by foster parents in a far country.

9) We are told [little or] nothing of his childhood, but

10) On reaching manhood he returns or goes to his future kingdom.

11) After a victory over the king and/or a giant, dragon, or wild beast,

12) He marries a princess, often the daughter of his predecessor, and

13) Becomes king.

14) For a time he reigns uneventfully, and

15) Prescribes laws, but

16) Later he loses favor with the gods and/or his subjects, and

17) Is driven from the throne and city, after which

18) He meets with a mysterious death,

19) Often at the top of a hill.

20) His children, if any, do not succeed him.

21) His body is not buried, but nevertheless

22) He has one or more holy sepulchers.

Without going into the sordid details of how our favored real and mythological heroes actually end up conforming to Raglan's grand plan, we may still ask what it is about someone who achieves at great risk something physical that entitles that person to lead people—even if to lead them very badly. Is it possibly that we value the heroic paradigm—in politics, in film, in our advocacy work, in (especially in) anthropology—even more than life itself?

When citizens alarmingly sacrifice themselves to a favored myth, intelligent people set to work. Some, like Charles Lindholm, focus on the charismatic abilities of leaders to embody collective ideas. Others, like Elias Cannetti, tell us that such heroes are largely driven by the paranoid idea that there is hardly anything on earth worse than their own death—how they are driven to achieve by the fear of dying before they have been properly applauded; how an early death would strip them of the experience of their own glory.

At the same time, we all know (at least Raglan did) that most who take this route end up at some point being caught without their clothes—that Arnold the Terminator will almost surely be publicly humiliated at some time in the future by his behavior, or that Messner will be haunted by former friends and colleagues who now argue that he abandoned his own brother on a climb that went so wrong that his own flesh and blood (like an off-spring of Herakles) was, so they claim, sacrificed to his own heroic needs.

Somehow, in other words, we all sense that these outcomes will happen, but that doesn't stop most would-be heroes from going for it anyway. Is it for the reasons suggested by Lindholm, Canetti, and others? Does "Tarzan" strive to become a king or peacemaker because he possesses certain skills, or because some of us cannot live without the thought that there are still a couple of people on earth for whom the visceral catalysts of risk can be resolved in a way that allows the brave of heart to change themselves and those around them? In that case, heroic narrative steals the anonymous embodiment that might have satisfied the most adventuresome and forces the adventurer's trials to be acknowledged.

Perhaps, then, it is the final fear of dying anonymously—without having left one's mark—that, more than any other fear, provides the catalyst for writing; for it takes little imagination to grasp how the authority obtained through representing the weak especially resembles the self-aggrandizing authority of those paranoid tyrants of old (and anthropologists of present) who, to rephrase Canetti (1960), could not imagine anything so horrific as their own premature demise. Indeed, because of the subtle shift from a genuine experience of pain to the authorial appropriation of another's suffering, the anthropological literature on suffering—a.k.a. illness experience—perhaps even more so than that of adventure—becomes vulnerable to vicarious self-promotion. This is what Stoller means when he (more graciously than I) advises scholars to adopt the "respectfully decentered conception and practice of depicting social life" (1997, 26) that characterizes the outlook of West African storytellers—those "griots" who "are humbled by history, which consumes the bodies of those who attempt to talk it, write it, or film it" (ibid.).

Like Thesiger's ambivalence about an adventurer's links to civilization, Stoller's West African storytellers remain very wary of their audience's tendency to conflate a storyteller's own person with the epic he has embodied. The true griot, in other words, would never allow his audience to see his accomplishments as an outcome of his greatness. Such foolishness, Stoller's sages believe, is reserved for authors; for the real adventurer, like the true storyteller, must remain enough decentered to give over the epic his audiences yearn for without giving over the private embodiments that are his alone. To offer these in exchange for easy flattery from the uninitiated, would be to allow oneself to be falsely deified. As Raglan showed so admirably (so long ago), to do this would ultimately be to walk foolishly in the direction of one's own crucifixion. Perhaps, in the end then, it's knowing that one could never imagine the private miseries and ecstasies of Edgar Rice Burroughs's *Tarzan of the Apes* that makes so enduring an experience of imagining what they might be.

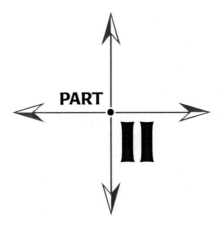

Exhibitionary Adventures

CHAPTER 6

They Sold Adventure: Martin and Osa Johnson in the New Hebrides

Lamont Lindstrom

The twentieth century dawns and the high modernist crash of urbanism, industrialism, bureaucracy, new mass media, and the centralizing state gives young sociologists something to write about. How can society and the individual coexist? In France, Émile Durkheim worries about social stability and community given disparate, often conflicting individual interests. The division of labor, luckily, provides some moral glue. Across the Rhine, Georg Simmel picks up the German end of the stick. He frets that the tightening imperatives of mass society have corroded and subjugated humanity. In response, he produces a series of optimistic portraits of individuals who manage to live in society yet nonetheless retain a golden measure of autonomy and authenticity. Not everyone is a mere cog in the societal works. There are also the Stranger (see Levine 1979), the Renegade, the Noble, the Artist, the Gambler, and the Adventurer.

The adventurer, in particular, bravely steps beyond his social estate to compose his own life. He "makes a system of life out of his life's lack of system" (Simmel 1911: 191). He rises above the crowd and, not bogged down in life's substance, instead experiences life's process (1911: 198). Simmel's model romantic adventurer was Giacomo Casanova. A sometime musician, diplomat, confidant, secretary, and spy, Casanova was deeply involved in European political and religious circles of his day. But he was also significantly distanced from these. He was exiled while a young man from his home, Venice, and throughout the rest of his life he moved from city to city across the continent. Casanova's infamous autobiography, which covers only his first forty-nine years, chronicles his adventures with more than one hundred women. Thus estranged from European norms of family, marriage, and citizenship, Casanova had adventures. And so, Simmel hoped, might anyone who could find his way into "something alien, untouchable, out of the ordinary" (1911: 189).

But, too late, others then were also longing for adventure. Simmel argued that an adventure can seem like a dream insofar as it is difficult to assimilate. And indeed, he continued, "we may think of it *as something experienced by another person*" (1911: 188, my italics). He then moved along to discuss the character of the authentic adventurer who is like an artist, a gambler, a genius, a lover, a conqueror, and so forth. Simmel's passing remark about the detachability of adventure, however, pointed to a commodification that was already then underway. If your own adventures sometimes seem like those of another, then perhaps somebody else's adventures could feel like your own. My guess is that Simmel would have been suspicious of vicarious, fake adventuring as a symptom of, rather than antidote to, modernity. It is better to be Casanova than to read his memoirs. Others, however, better appreciated and were beginning to cultivate the growing market for adventure stories. The dime novel had already appeared in the mid-nineteenth century and, fifty years later, the early motion picture industry was building on its cheap genres of romance, comedy, melodrama, and adventure (Enstad 1999: 1972). Among these fictions were adventure stories sold by real adventurers; and these first-person accounts often commanded greater returns in the developing mass market for vicarious adventure.

In 1907, four years before Simmel extolled the purer merits of adventuring, Jack London sailed from the Oakland Estuary on his fifty-seven-foot yacht, the *Snark*. During the previous decade, London had become famous as an American entrepreneur of adventure. His relations with adventure were complex. He wrote fictional accounts of made-up adventure. Some of these, however, were based on his own exploits and he also published autobiography, essays, and travelogues. By 1907, London could have retired to his newly built Valley of the Moon mansion, writing up potboilers and tearjerkers at his comfortable desk. Instead, with his second wife, Charmian, he sailed through the Golden Gate and headed for Hawai'i. He sought adventure. But he also sought money.

On board the *Snark* was a twenty-two-year-old Kansan named Martin Johnson. The year before, Johnson had read one of London's publicity pieces for the upcoming voyage in his sister's copy of *Women's Home Companion* (Imperato and Imperato 1992: 25). As a publicity stunt, the Londons proposed to take on board one volunteer, and Johnson wrote offering his services, particularly with photography. Picking out Johnson's letter from a considerable pile of other applications (1992: 27), the Londons hired him as their ship's cook. Johnson made his way to Oakland, California, and, after several months of delay, embarked on what all thought would be a round-the-world cruise. After a taxing voyage to Hawai'i, the *Snark* sailed south to the Marquesas, Tahiti, and Raiatea, and then east to Samoa, Fiji, and the New Hebrides (today's Vanuatu).

While en route to Hawai'i, London busied himself writing a magazine piece entitled "Adventure" (C. London 1915: 18). Ultimately, the adventures and misadventures of the *Snark* became well-known in that the Londons wrote about these, both during the voyage and afterwards (e.g., Jack's *The Cruise of the Snark* (1911), and Charmian's *The Log of the Snark* (1915). Martin Johnson, in fact, surprised and angered Jack and Charmian by publishing in 1913 his own account of the voyage, *Through the South Seas with Jack London* (ghostwritten by Ralph D. Harrison), before Charmian had finished her own book (Imperato and Imperato 1992: 62). Copying Jack, Johnson's first chapter is titled "On the Trail of Adventure."

Of the original crew, Johnson was the only one still aboard with the Londons when the expedition finally ended up in the Solomon Islands in November, 1908. All of the crew was ill. Jack, suffering from pellagra and malaria, Charmian, and Martin all embarked on the *Makambo,* a Burns Philp steamer, for Sydney where they laid up to recuperate for several months (Imperato and Imperato 1992: 49). London, worried about his health, especially his swollen hands, decided to end the voyage. Johnson caught a steamer back up to the Solomons and, with its crew, brought the *Snark* down to Sydney where the Londons sold it. They then returned to California while Johnson, continuing westward, booked passage home to the U.S. via Europe.[1]

Broke, Martin returned to Kansas in 1909. He owed the Londons $487 and he needed a job (Imperato and Imperato 1992: 52). Martin was not much interested in joining his family's business, a jewelry store in Independence, Kansas, despite his father's offer to invest in an expanded photography studio. Before the *Snark,* he had explored other avenues of escape from small-town Kansas, including working as a roustabout for the Forepaugh-Sells Circus and as a stockman at the Zack Mulhall Ranch near Guthrie,

Oklahoma (Imperato and Imperato 1992: 16). Zack Mulhall, like Jack London, was another early adventure entrepreneur who repackaged the Western cowboy for popular consumption. In 1899, he had formed "the Congress of Rough Riders and Ropers"—one of the original Wild West shows in which Will Rogers and Tom Mix both performed. His daughter, Lucille, also rode in the show and billed herself as "America's First Cowgirl" (Day 1955). Johnson had traveled to Europe, accompanying two lots of Mulhall's horses and mules being shipped to England—the stock of Wild West shows on tour (Imperato and Imperato 1992: 16).

But, now back home again in Kansas, what could Johnson do? Martin decided to emulate Jack London and Zack Mulhall, who were both making a decent living from selling adventure, and themselves, as South Seas Explorer and Western Cowboy. Johnson, now a local small-town hero, went into business with an Independence druggist who converted his store into the *Snark* theater. On its stage, Johnson began his career as a purveyor of adventure, presenting travelogue lectures illustrated with South Seas photographs (Imperato and Imperato 1992: 52). Johnson and the Londons had taken many of these images, and he bought additional Pacific slides from several other sources.

Onstage, Martin billed himself as the "travelogue" man (Imperato and Imperato 1992: 53), borrowing the term from Burton Holmes, who had coined the word a few years earlier in 1904 (1992: 250). Holmes was a popular American photographer and lecturer who also was making a living by giving illustrated lectures about Pueblo Indians, the Klondike, Japan, and so forth. These travelogues sold voyeuristic adventure to stay-at-homes, in Kansas and elsewhere—sedentary folk who would only begin going overseas themselves in the 1960s with the advent of mass tourism.

The public travelogue, a marketplace creation of vicarious adventuring, emerged at this point thanks to the invention of several key technologies of modern mass media. Notably four developments in the latter years of the nineeenth century made possible Johnson's commercial adventurism. These were dry plate photography, electric arc lamps, halftone photomechanical printing, and especially moving pictures.

George Eastman had introduced in the latter years of the nineteenth century dry glass photographic plates, flexible film rolls, and eventually the Kodak camera, which hit the market in 1888. To create a mass market for cameras, Eastman distributed Kodaks, gelled plates, photographic paper, and film in department stores, drugstores, camera shops, and also in jewelry stores such as the one in Independence, Kansas, owned by Martin's father (Phillips 1996). The Imperatos, Johnson's biographers, suggest that Jack London chose Johnson from among the hundreds of others wanting to sail on the *Snark* because of his previous experience with cameras. He knew that, by 1906, the public was demanding that stories of adventures be illustrated.

Johnson, copying Burton Holmes, illustrated his travelogue lectures with photographic, often hand-colored, glass slides projected in magic lanterns. Although slide projection techniques date back to at least the mid 1600s, it wasn't until the invention of oxyhydrogen limelight and, by the end of the nineteenth century, the electric arc lamp that public presentations to large audiences became possible.

Adventure imagery also sold in print form as well as in public lecture. Editors sought out London and other adventure writers to fill magazine pages with illustrated adventure. The same year the Kodak camera went on sale, Frederick Ives developed the crossline halftone screen (Phillips 1996). Halftone technology made possible for the first time a mass production of photographic imagery. In the 1890s, a flush of "ten-cent illustrated magazines" appeared featuring halftone photographs, including *Cosmopolitan* and the *Women's Home Companion* that partly had supported the cruise of the *Snark*. London had offered, unsuccessfully, to name his yacht the *Cosmopolitan Magazine* if that magazine would agree to cover costs of his voyage in full.

Simmel circumscribed his definition of adventure with a frame of similes. Adventure is like a dream, a gamble, a love affair and, especially, adventure is like a work of art. The essence of art (and adventure) is that

> it cuts out a piece of the endlessly continuous sequences of perceived experience, detaching it from all connections with one side or the other, giving it a self-sufficient form as though defined and held together by an inner core ... it is an attribute of this form to make us feel that in both the work of art and the adventure the whole of life is somehow comprehended and consummated (1911: 189).

Simmel probably had in mind the connoisseur losing himself in a painting or the reader sinking into her novel. The frame of a painting or the covers of a book are like the bounds of an adventure, cutting it away from everyday experience. But so were the new mass-produced halftone photos and the motion pictures that Simmel must also have noticed by 1911. The new cinema, in particular, invited viewers within the frame of darkened theaters to lose themselves in the adventures of others.

In 1909, when Johnson hit the travelogue circuit with his magic lantern slides, the public had already begun to demand motion pictures. The Lumière brothers had exhibited the first commercial cinematic production in 1895, selling their business to Pathé Frères in 1900. Before the establishment of the Hollywood studios, Pathé Frères in Paris was the largest film production company in the world, supplying moving pictures for nickelodeons and, subsequently, theaters. The Londons and Johnson had, in fact, met up with an ailing Pathé Frères film crew in 1908 on Guadalcanal (Solomon Islands) that was hunting for cannibals. A plantation worker guided the filmmakers to his village, a few miles inland, and Johnson tagged along to help film the hoped-for man-eaters (Imperato and Imperato 1992: 48).

Passing through Paris on his way home to Kansas in 1909, Johnson purchased a reel of this film from Pathé and this would add several minutes of motion picture to his lecture shtick (Imperato and Imperato 1992: 56). Johnson also leaned on the Londons to lend him more of their *Snark* photographs and he eventually purchased additional South Seas films. After a year on the road, performing in small Midwestern towns, Johnson signed with agents of the Sullivan-Consedine and then the Orpheum vaudeville circuits. The latter booked his "Through the South Seas" travelogues in markets as big as London and New York.

In the slow summer season of 1914 when theaters closed down because of the heat, Orpheum hired Johnson to film short travelogues around New York City and Boston (Imperato and Imperato 1992: 63). Three years later, drawing on this film production experience, Johnson convinced a group of Boston investors to fund a photographic expedition back to the New Hebrides and Solomon Islands. Like the Pathé crew he had encountered on Guadalcanal ten years previously, he was on the hunt for cannibals.

This time he brought along Osa, a Kansas girl he had married in 1910 and who, since their marriage, had traveled with him on the vaudeville circuit. In the Midwest and Canada, Osa had sung and performed improvised Hawaiian hula dances to accompany Johnson's slides and lectures (photo 6.1). This proved too cornball for more sophisticated audiences, and when the act moved east she just worked the magic lantern (Imperato and Imperato 1992: 59). But when the Johnsons reached Melanesia in 1917, Martin put Osa back on stage by placing her in front of his camera.

Simmel had thought women, in general, too passive to be adventurous. Adventures—love affairs in particular—require a tricky combination of

> conquering force and unextortable concession, winning by one's own abilities and dependence on the luck which something incalculable outside ourselves bestows on us. A degree of balance between these forces, gained by virtue of his sense of their sharp differentiation, can, perhaps be found only in the man (1911: 195).

Women, along these lines, fail to share the adventurer's delicious sensation of being in control of life while at the same time allowing himself to be carried along by life. Women are too used to be carried along in normal, nonadventurous moments to appreciate submissiveness in special times. A woman is too permeated by passivity "which either nature or history has imparted to her character" (Simmel 1911: 195).

Figure 6.1 • Osa in hula dress.
(courtesy Martin and Osa Johnson Safari Museum)

Nonetheless, even though Osa (billed as Mrs. Martin Johnson) played second fiddle to her even more adventurous husband, her adventures were on sale too. Johnson recognized a market for female adventure stories. He was not the first in this. Obviously, as the Imperatos note, he copied the Londons. Osa was to be his Charmian. Johnson may also have had in mind the cowgirl Lucille Mulhall (daughter of his former boss) along with other then-popular female Wild West show stars (e.g., Annie Oakley, Calamity Jane). Movie producers had also recently introduced several women's adventure serials, including *What Happened to Mary,* beginning in 1912, and *Hazards of Helen* in 1914 (Enstad 1999: 172, 193). And, also in 1914, Edgar Rice Burroughs introduced his remarkably popular adventurous couple, Tarzan and Jane, in *Tarzan of the Apes.*

The genealogy of adventuring women, of course, traces back to British women making the Grand Tour of eighteenth-century Europe, and to the even more daring women travelers of the Victorian age (Dolan 2001). The intrepid Beatrice Grimshaw, Irish sportswoman, traveler, and writer, passed through the New Hebrides about the same time as the *Snark*. These female tourists had published a variety of accounts of their travels, more or less known (Dolan 2001: 287). At the beginning of the twentieth century, illustrated magazines and moving pictures, many directed toward a female audience, had much expanded the market for women's adventure stories even further and Johnson was determined to profit from this.

The Johnsons landed back in Melanesia in 1917 with two still cameras, one motion picture camera, and forty-thousand feet of unexposed movie film. They were in the New Hebrides between October and December, mostly in Port Vila (Efate) and Vao, a small island off the coast of northeast Malakula. They shot nearly twenty-thousand feet of film during these three months (Imperato and Imperato 1992: 73). In 1919, with new financial backers, they returned to the New Hebrides for several additional months of filming, coming back into Sydney with another twenty-five thousand feet of exposed footage. From there, they went west to Borneo, and eventually on to Africa and to successful careers as wildlife photographers, filmmakers, safari guides, and (in the case of Osa) early television personality. Their work sold fairly well in the emerging adventure marketplace. After 1919, the Johnsons lived successfully off adventure; no further recuperative trips back to Kansas were necessary. Martin was killed in a California plane crash in 1937 and Osa, after taking some of their films onto TV in the early 1950s, died of a heart attack in 1953.

On both trips to the New Hebrides, the Johnsons set up headquarters on Vao where there was a French Roman Catholic mission station. (Several years earlier, anthropologist John Layard had spent part of 1914 and 1915 on the islet.) During their first expedition, Johnson arranged for a ship to take them around from Vao to Tenmaru on Malakula's northwest coast. Here, on the beach, they met their first Big Nambas[2] man who approached them complaining of stomach pains. Osa, recounting the meeting in *I Married Adventure,* one of her ghosted memoirs, stoked up the savagery:

> He was the most horrible looking creature I had ever laid eyes on. Coal black and incredibly filthy, his shock of greasy hair and heavy wool beard were probably the nesting place of every sort of vermin. A gorget of pig's teeth hung around his neck; he wore a bone through his nose and he was entirely naked except for a large breach clout of dried pandanus fiber. As he came nearer I saw that his deeply creviced face was horribly distorted. It made me think of a grotesque mask (O. Johnson 1940: 117).

The Johnsons made friendly, however, by giving him a few cascara pills—a popular laxative of the day. Learning from their new beach acquaintances that Nagapate,[3] who they presumed to be the "chief" of the Big Nambas, was living nearby, they struck out inland to capture him on film.

In Nagapate the Johnsons literally found their poster boy (photo 6.2). Back in New York, they focused a good deal of their literary and cinematic South Seas productions around his image. Nagapate's "every gesture was chiefly," wrote Martin (1922: 17):

Figure 6.2 • Nagapate.
(courtesy Martin and Osa Johnson Safari Museum)

he was enormously tall, and his powerful muscles rippled under his skin, glossy in the sunlight. He was very black ... his expression showed strong will and the cunning and brutal power of a predatory animal.... On his fingers were four gold rings that could only have come from the hands of his victims (1922: 17).

And, more than Nagapate, Martin would eventually claim all the Big Nambas for himself:

In a sense, they were my people. They had encircled the globe with me and in the comfortable surroundings of great theaters had stood naked and terrible before thousands of civilized people. I had made their faces familiar in all parts of the world (1922: 61).

Martin and Osa puffed their adventurous selves by boosting Nagapate's conjoint nobility and savagery, along with the terrible nakedness of all his tribe.

Returning again to Vao in 1919, the Johnsons once again sought out Nagapate. This time, they had also brought along a portable generator, a film projector, and a screen that they set up on the beach at Tenmaru to show the Big Nambas some of Johnson's 1917 footage in which they had starred (photo 6.3). While the projector rolled, the Johnsons filmed a bemused Big Nambas audience lit up with radium flares. On their return again to the U.S., they featured these scenes in new cinematic productions as "cannibals at the movies" (Imperato and Imperato 1992: 80). Harald Prins has noted that this is the earliest known instance of "cinematic feedback" (1992: 9), where a filmmaker films subjects watching themselves on film.[4]

The Johnsons' strategy was not just to photograph primitive natives but also to photograph themselves dealing adventurously with savages (photo 6.4). The Imperatos write that "they wanted movies that presented the South Pacific islanders as wild creatures whose lives were governed by a terrifying yet fascinating irrationality" (1992: 6). Savagery is efficiently defined and foregrounded when juxtaposed with civilization (this is, of course, the self/other binary of much late theoretical comment), and the Imperatos go on to say, "the combination of the exotic, dangerous savages, and courageous Midwesterners filled the bill perfectly" (1992: 6). Audi-

Figure 6.3 • Examining the gear.
(courtesy Martin and Osa Johnson Safari Museum)

Figure 6.4 • Martin with two men from Vao.
(courtesy Martin and Osa Johnson Safari Museum)

ences saw imagery of irrational and treacherous New Hebrideans mixed up with a dose of Kansas common sense. Martin and Osa, their cameras running, acted out adventure. And, as the Imperatos note, for the rest of their lives they "played the parts they first created for themselves in their New Hebrides adventure" (1992: 6).

Acting adventurous, for the Johnsons, involved playing with a series of savage/civilized juxtapositions. By 1917, this sort of contrast was nothing new. Putting oneself in close contact with possibly dangerous and different others embodied Simmel's basic definition of adventure, whose most general form "is its dropping out of the continuity of life" (Simmel 1911: 187). The voyage of the *Snark* was discontinuous in this way from ordinary, non-adventurous life back in Kansas or even California. Jack London, in fact, had posed Johnson while on Guadalcanal in 1908 in a striking photographic juxtaposition (photo 6.5). In this image, Martin culturally and racially cross-dresses (or perhaps undresses), standing next to a young Guadalcanal man who holds bow and arrows. Martin crosses his arms defensively, though his gesture probably responds more to his own nudity than to the armed native at his side. The photographic image, however, frames an adventure: Simmel's comprehensive inner core romantically detached from everyday experience.

We might have a closer look at the parts that the Johnsons "first created for themselves" during their New Hebrides adventure. Although films were their big moneymakers, the Johnsons also continued to take still photographs, many of which they published in their various books and articles (e.g., M. Johnson 1913, 1922; O. Johnson 1940, 1944). Since Martin was typically behind the camera, we might look especially at his poses of adventurous Osa—Jane to his Tarzan. As the Imperatos note:

> Martin concluded that Osa's true value lay in front of the camera's objective, and not behind the viewfinder. She eagerly posed and pantomimed with the islanders, and the resulting screen images of a petite flaxen-haired American woman surrounded by cannibals and headhunters later jarred audiences and drew them to the box office (1992: 70).

Figure 6.5 • Martin cross-dressed.
(courtesy California State Parks, 2005)

During their first, nervous meeting with Nagapate in 1917, Martin pro-
pelled Osa forward and began cranking the camera: "'Remember, darling,'
his voice was low and quiet, 'show no fear—smile—open up the trade stuff'"
(O. Johnson 1940: 120).

Johnson exploited the power of the camera to create adventures of the
familiar with the strange, and to freeze these images so that he could sell
them back home. White stands next to black; civilized next to savage. He
imported from Kansas much of the language of American racism. His writings found the New Hebrides to be full of bucks, big slobbery lips, and
black slaveys. Johnson had soaked up pop evolutionary theory of the day,
and his Islanders were therefore primitive, savage, orgiastic, unpredictable,
childlike, and—so he hoped—cannibalistic. He posed himself, and Osa, as
civilized white in contrast and in contact with savage black native. In one
shot, Osa's stance echoes that of a Malakulan man standing at her left, but
she is dressed in gingham and he in penis wrapper, she in boots, while he
is barefooted (photo 6.6). Osa stares at what he holds: a kinsman's mod-

Figure 6.6 • Osa and Tomman men.
(courtesy Martin and Osa Johnson Safari Museum)

eled head, preserved during local funerary ritual. Johnson published another photograph of cured heads in his 1922 book, titled "The Old Head-Curer." When Osa republished this in 1940, this photo had become the more obviously savage "Campfire of the Headhunters."

Martin also came out around to the front of the camera to share the frame with his subjects. Nowadays, this would be called a reflexive move but, as Prins has noted, sharing the frame may carry several messages about the quality of a photographer's relation with his subject (1992: 14). Within a shared frame, the self may display himself as superior to the other, as inferior, or as an equal. Much of the Johnsons' work celebrated the first of these stances but, for purposes of selling adventure, it did not always pay to be too superior. The savage's larger numbers, or greater strengths, or jungle skills, or local know-how might always turn the tables. Johnson staged one photo contrasting his six-foot height with shorter Islanders from Espiritu Santo (Pygmies, he called them) (photo 6.7). They are armed and he is outnumbered, although his foregrounded rifle evens the score.

The adventure marketplace understandably induced the Johnsons to accentuate savagery and its dangers.[5] Martin had returned to the New Hebrides in 1917 after finding the Solomon Islands to be too civilized. Arriv-

Figure 6.7 • Martin and "Pygmies."
(courtesy Martin and Osa Johnson Safari Museum)

ing on Vao, he was on the hunt for untamed savages, the paramount of which would be cannibals. Johnson figured that pictures of cannibals would sell easily, particularly footage of cannibals in action (Scheinman 2000). He had, however, turned up his nose at staging a cannibal feast or two when in the Solomon Islands. He "was a patient, persistent artist who would never be satisfied with anything but the truth" (O. Johnson 1940: 112), or so Osa claimed.[6] And on Malakula he found that truth. Johnson's book, *Cannibal-Land,* is structured as a successful quest. The story culminates as Martin, lugging his motion picture camera, sneaks up on a twilight cannibal feast. He shoots this at a distance through a telephoto lens before sending an assistant in to drop a flare into the fire. This, understandably, frightens people who grab their dinner—that precious evidence of cannibalism—and run off. Johnson wrote triumphantly: "I had proved what I had set out to prove—that cannibalism is still practiced in the South Seas. I was so happy that I yelled" (1922: 189). The Imperatos (1992: 82) conclude that Johnson actually had filmed yet another bout of funerary ritual head curing but, no matter, this would play as cannibalism in the world's theaters.

The savages, however, could sometimes be too savage. Johnson was particularly vexed by penis wrappers, then worn on several New Hebrides islands. He had first encountered these on Tanna in 1908, where Jack London described local men as "worse than naked" (M. Johnson 1913: 277), presumably because wrappers hold the penis in erect position. Johnson complained, "The hundred or more pictures I made of these cannibals cannot be printed in a volume intended for popular circulation" (1913: 268). (He did, however, show some of these while on the vaudeville circuit during special "men only" lectures (Imperato and Imperato 1992: 59).) Johnson also had problems with his 1917 Big Nambas footage, writing:

> the dress of the men of Malekula, if you can call it dress, calls attention to their sex rather than conceals it. On my first visit among them, I had taken motion-pictures of them as they were. When I returned to America, I found that naked savages shocked the public. Some of my best films were absolutely unsalable (1922: 181).

During his 1919 expedition, Johnson did his best to cover up his cannibals with "geestrings or loin-cloths or aprons of leaves" (1922: 181), as is apparent in a shot he published of teenaged boys on Vao. Cannibalism yes, and bare breasts certainly, but American audiences were not yet ready for savage testicles, adventure or no.

Alongside the simple civilized/savage juxtapositions that fueled the Johnsons' adventures, Martin and Osa sometimes also posed themselves as *like*
the natives, at least up to a point. Johnson even struck occasional notes of
relativism. For example, he made fun of small-town Kansas businessmen's
lodges as an American sort of New Hebrides's graded men's societies (M. Johnson 1913: 158). Osa, in particular, mediated the civilized/savage opposition.
As female, she was closer to nature, to animals, to motherhood, and to children—including big childlike savages like the Big Nambas. Later in her career, she specialized in the softer side of the adventuring business, writing a
series of children's books such as *Osa Johnson's Jungle Friends, Jungle Babies,*
and *Jungle Pets.* Martin often posed Osa alongside women and children. In
one photo, she sits in a row with a number of Big Nampas women—and
like them she squats low and covers her head (photo 6.8). And Osa's sexuality could also create useful danger for the camera. The Johnsons suggested that Nagapate, attracted to her white skin and golden hair, lusted to
add her to his harem of wives (M. Johnson 1913: 88–89)—this the Big Ape
and White Girl story that Hollywood's *King Kong* would popularize again
in 1933.

Figure 6.8 • Osa and Big Nambas women.
(courtesy Martin and Osa Johnson Safari Museum)

Simmel may or may not have counted savage kidnap and rape, or even cannibal feasting, to be adventurous in his purer terms. They played well, however, in theaters then springing up everywhere. By 1911 it was difficult to be a Casanova. The romantic individual, at the acme of high modernity, was about as dead as the dodo. But if it was tough anymore to have one's own personal adventures, these could now, thanks to emerging mass media, be purchased and vicariously experienced in book, magazine, and theater. Martin and Osa Johnson were among the first entrepreneurs to grab camera, head out for some cannibals, and capture adventure on film. But this was adventure of a remarkable and marketable sort. While Casanova, Simmel argued, was during his adventures "entirely dominated by the feeling of the present" (1911: 190), the Johnsons, during theirs, lived under the eye of the camera and its future returns. The value of adventure now rested in the images waiting, in the pregnant womb of the camera, for future development. Martin and Osa Johnson may have married adventure; but they sold it too.

Notes

1. Before leaving Australia, Johnson wrote to Theodore Roosevelt volunteering to join a South African safari and scientific collecting expedition that Roosevelt was leading. His attempts to attach himself to this second celebrity adventurer failed (M. Johnson 1913: 368).

2. Throughout their writings, both Johnsons, probably for marketing purposes, persisted in calling these people the "Big Numbers." *Namba* is the Bislama (Vanuatu Pidgin English) term for penis wrapper—an object that Johnson needed photographically to erase. Although Australian pronunciations of "numbers" and "nambas" are near homonyms, and perhaps misleading to some American ears, Johnson knew the term *namba* and defined this correctly in his book about the 1919 expedition (M. Johnson 1922: 9).

3. The name actually was Nihapat according to anthropologist Kirk Huffman (Imperato and Imperato 1992: 240, n. 11).

4. Johnson, years earlier, had shown Solomon Islanders pictures that he took of them during the *Snark* cruise (M. Johnson 1913: 326). While in Pape'ete, he had also observed the emotional impact of early cinema on a Tahitian audience: "... magic pictures that none of these people had ever seen the like of. It was many a day before the natives could understand that it was not supernatural" (1913: 207).

5. Edward Curtis's silent film *In the Land of the Head-Hunters* (1914) had at-tracted a large audience and the Johnsons may have had this in mind.

6. The Wild West shows, with which Johnson was acquainted, also advertised their authenticity—their casts of "real" cowboys and Indians (Browder 2000: 61). Browder suggests that the market value of authenticity responded to growing sus-picions of urban modernity wherein people more easily pass for what they are not (2000: 54).

Chapter 7

Jacaré: Cold War Warrior
from the Jungles of the Amazon

Neil L. Whitehead

As Rodney Needham (1983) indicated in his inspiring consideration of the literary figure of Tarzan, the realm of the fictive and imagined is a significant site for the appreciation of cultural practice and proclivity. As Needham rightly emphasizes, Tarzan-like constructions may be understood as offering the vision of premodern freedom that liberates the conventional and rule-bound existence of the "civilized." As a result many such fictions pre-date the Tarzan figure: Rousseau's "noble savage," and its variants, the interest in wolf-children, or even the somewhat bourgeois existence of Daniel Defoe's Robinson Crusoe, also speak to this fascination with an escape from the mundane and ordered world of modern society.

In order to appreciate some of the further meanings of such literary devices in this chapter I want to introduce the figure of Jacaré,[1] the fictional hero of Victor G.C. Norwood's "Scion Jungle Novels" serialized in five works, beginning with *The Untamed #1*. Jacaré first emerges from the jungles of Brazil and Guyana in this 1951 yarn and goes on in a subsequent adventure, *The Skull of Kanaima #4* (see fig. 7.1), which is the focus of consideration here, to fight the good fight against evil Russian agents seeking plutonium supplies in the uplands of northern Amazonia.

Jacaré also battled more supernatural and monstrous foes in other adventures such as *The Caves of Death #2* (see fig. 7.2) where the hero discovers a subterranean "Lost World," an inversion of Arthur Conan Doyle's mountainous locale, which actually serves to emphasize rather than distract from

Figure 7.1 • *The Skull of Kanaima* (London: Scion Ltd., 1951).
Every effort has been made to trace the copyright holder and to obtain permission for use of this image.

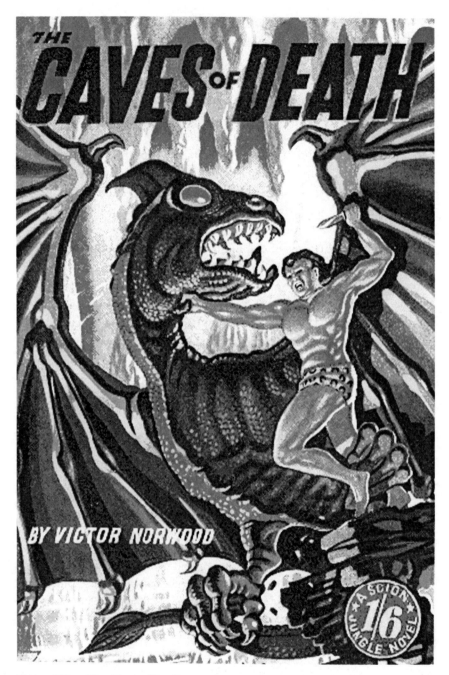

Figure 7.2 • *The Caves of Death* (London: Scion Ltd., 1951).
Every effort has been made to trace the copyright holder and to obtain permission for use of this image.

the plagiarism. The episode illustrated on the cover pits Jacaré against a fierce pterodactyl, and is itself an iconic episode in Arthur Conan Doyle's original work. In *The Island of Creeping Death #5*, Jacaré is shown on the cover in a scene closely identified with Johnny Weissmuller's crocodile-wrestling Tarzan scenes, as well as with Roy Rockwood's *Bomba the Jungle Boy* series.

As well as this obvious hypermasculinity, Jacaré can also represent innocent and untutored savagery, which, nonetheless, is to be favorably compared with dissolute and primitive barbarism, as appears in his battles with a native priestly sect in *The Temple of the Dead #3*. For the cover Jacaré is therefore shown performing a suitably Christ-like ejection of the priests from said Temple. Such a motif was also expressive, as will be discussed further below, of the battle of the good and honest violence of democracy against the evil seductions of communist tyranny.

The Jacaré series of books by Victor Norwood (1920–1983) was clearly a commercial attempt to ride the wave of the popularity of the "Tarzan" figure, but here transplanted and reimagined in an Amazonian setting. In these terms alone the Jacaré series is not necessarily the most complex or completely realized Tarzan clone since there are numerous examples of other kinds of Tarzanesque characters from this period. The reason that the Jacaré series holds particular interest for an anthropology of adventure is that Norwood himself "adventured" in British Guiana in the 1940s and tried his hand at diamond mining there. The publisher's blurb on the inside cover of his account of this adventure is worth quoting in full:

> Victor G. C. Norwood has not yet realized his ambition of traveling across Africa in an amphibious vehicle taking photographs on the journey. He has, however, crammed into his 42 years a great deal of action and adventure. Happily married with two sons aged 22 and 7, Mr. Norwood has traveled extensively in Africa, America and Europe and spent many years diamond prospecting in British Guiana and Brazil.
>
> He was a former heavyweight boxing and wrestling champion until he lost two fingers in a skirmish with Brazilian revolutionaries. During the last war he served in the Merchant Navy as Q.M. Machine Gunner until he received severe injuries due to enemy action. Despite his constant search for adventure in remote places, Victor Norwood had two years operatic voice training and has sung in all parts of the world.

Given this self-presentation, the focus of analysis here will be interplay between Norwood's self-imagination and the figure of Jacaré, the work of colonialism in the representations of Jacaré and his world, as well as the turn to modernist topics such as the Cold War and atomic war. In this way the type of materials produced by a post-WWII figure like Norwood will also be an opportunity to assess the wider meanings of "adventure writing" in the emergence of the postcolonial world.

The figure of Jacaré spans a series of five pulp novels but the figure of Norwood himself is highly interesting for the connections he suggests between the fictional and the adventurous, the imagined and the ethnographic, since all these elements are swirled around in a vivid, if decidedly anachronistic, tale of jungle derring-do. Before considering the author himself, as well as the milieu in which he moved, the figure of Jacaré needs first to be investigated.

I propose to consider in some depth just one of Victor Norwood's novels, *The Skull of Kanaima*. This approach is suggested by my own ethnographic interest in kanaimà, which is how I first encountered the Norwood novels, and thus also provides the opportunity to comment on the use of the device of ethnographic realism in the novel itself. Further analysis of the complete canon must await another opportunity.

The Skull of Kanaima

The formulaic nature of the Jacaré figure is best illustrated by considering the literary nature of that formula itself, which most certainly derives from the Tarzan series of Edgar Rice Burroughs. This is evident in any number of ways, as the illustrative material itself makes obvious, but the more interesting process involved in this plagiarism is the sometimes quite subtle recasting of Amazonia as African—although, given Norwood's racism throughout the text, the cultural differences of Amazonia and Africa are hardly of importance to the author.

Aside from Jacaré himself, the main protagonists of the novel are the American, Rocks O'Neill, and the Britisher, Jim Trent, presumably a projection of his own self-image. Together they are sent by the Brazilian and British governments to locate the source of a uranium deposit located on the territory of a fierce tribe that can only be parlayed with using the "skull of kanaima," an important tribal talisman. The skull itself is shrunken[2] to the size of a small orange on which inscriptions were made by a powerful "witch-doctor" thousands of years ago. The tribe believes it to be the skull of a demon.

The skull had been kept for "countless generations" until it was stolen by a curio hunter, whom they nevertheless killed, although the skull was lost. It was recently rediscovered by one Jan Emmet, who made a deal with the British and Brazilian government to pay Rocks and Jim $5,000 each to recover the skull and so locate the uranium with the help of the grateful tribe.

The scene now moves to the trading post of one Margon, a "fat, loutish German" who is sheltering Pamela, the daughter of a missionary whose station was attacked by the "Tarawaddies." They find Jan Emmet stabbed in the back, who is able to whisper the single dying word "kanaima." It transpires that two Soviet Russians, Monrov and Virosky, are also after the skull and it was they who had killed Emmet to retrieve it from him, as well as a map that indicates the uranium source is in the land of the "Quarawete" Indians.[3] But Pamela, fleeing the lustful advances of Margon, is seized by the Russians, who return her to the trading post and decide to team up with the German.

Happily, Pamela is freed by Rocks and Jim, who kill Margon, take the skull, and set the Russians adrift downstream in a canoe as they head upriver to the land of the Quarawete. As they do, the drums start up in classic adventure mode and sure enough Indians soon appear, led by a "prancing medicine man" wearing a belt of monkey tails and decorations of "all manner of shrunken heads" including monkey, rodent, and human. However, the skull "pleases them" and they agree to trade for the "rock that shines" and give Rocks and Jim a jade skull with ruby eyes as a talisman for whoever returns to trade with them.

Enter Jacaré, "muscled like a Greek god" and six foot eight inches tall! Although his tongue has been cut out, rendering the need for dialogue irrelevant, the brooding hero is accompanied by a female facilitator, the blond Helen. At this point the Russians return and ambush our heroes, who manage to grab Pamela and the talismanic skull. But Jim and Rocks are not to be cheated of their $5,000 by these "communists," for ideology is not part of the bluff-masculinity of our heroes, who need to be paid for their services to democracy. The enigmatic Jacaré tracks the Russians and kills them, but Jim Trent is captured by Indians and killed. Rocks and Jacaré also finish off four other Russians who come in search of their companions, while Helen and Pamela befriend each other. This happy, handsome, foursome finally splits up and Rocks and Pamela leave with the jade skull talisman, having cleared the way for uranium mining for the British and Brazilian governments. Jacaré returns to the forest with Helen.

Jacaré and Tarzan

How then does Jacaré match up to his African counterpart? Like Tarzan, he is physically impressive, but he lacks the hint of elite heredity that made Tarzan an accessible savage. Tarzan also possessed the capability of speech although it was initially only the ape-language. Jacaré is mute through unspecified violence that ripped out his tongue. This lack of speech, as well as his apparent capacities to kill, thus problematizes Jacaré's humanity in a similar way to that of Tarzan. The figure of Tarzan poses issues for the relation between ourselves and other primates, the basis of language as a claim for humanity, and the role of genetics in the production of moral and social character. In contrast, the far less well realized character of Jacaré is much more a savage backdrop to the actions of Rocks O'Neill and Jim Trent. In this sense Jacaré is hardly present in Norwood's jungle novels, and the invocation of the Tarzanic motif seems far more opportunistically connected to the commerce of pulp novels than the literary qualities of the stories themselves.

The figure of Jacaré thus relies on Norwood's readership having some kind of passive awareness of Tarzan since the extreme lack of any character development, as far as Jacaré is concerned, means that readers are left to fill in the gaps for themselves. The Tarzan trope thus allowed Norwood to rely on his audience's presumptions that "jungle heroes" are known for their lack of speech, physical endowment, violent propensity, and emotive identification with the democratic rather than the tyrannic.[4] This is not to suggest that the presentation of other characters in the novel is any more convincing or profound—quite the contrary—but the comparison with Tarzan highlights what is important to the cultural meaning of that figure, as much as it reveals the basis for the verisimilitude of Norwood's own version, Jacaré.

Perhaps surprisingly, Norwood does not directly play off his own experience of Amazonia at all, which he otherwise recycled a number of times (Norwood 1956, 1964, 1974), as this might have been a way of establishing difference with the Burroughs canon and the authenticity of Jacaré himself. Indeed, quite the opposite is true, in the sense that Norwood heavily

Africanizes the Amazon context; witch doctors not shamans, jungle drums in place of war cries, chewing betel nut rather than coca, and so forth. Even the cover illustration to *The Skull of Kanaima* shows a Eurasian tiger, not an American jaguar. Clearly we have traveled to the realm of movie-land where the icons and symbols of the jungle scene have become hyperabbreviated to the point that they can now index the "tribal" as a supracultural condition. Indeed, it seems from the treatment of both plot and location in the text itself, that the intent was not to challenge movie producers with a new canon of referents but to show how from such existing referents the production of a film version of the Amazonian Jacaré could be easily effected. As a result, the text is highly visual in that it narrates the scene as "scenery," while the description itself moves around such scenery like the lens of a camera. The cinematic pretensions of Norwood were no doubt influenced, and probably directly suggested, by the making of the film *Jacaré* in 1941. The movie starred Frank Buck, a minor actor with a number of other lead credits to his name.[5] The unrated film had a running time of fifty-seven minutes and as such was a classic "B" movie feature format designed for the Saturday morning cinema.[6] It seems likely that the name of Norwood's "Tarzan" was itself simply taken from the film to obviate these possibilities.

However, the borrowings from the Tarzan canon are in fact rather superficial, restricted mostly to cover visualizations and a literary catalogue of scene-setting tropes. Although there is no way in which the content of the Jacaré novels could be construed as the literary equal to those of Tarzan, their evident popularity still makes them a credible source of cultural understanding of the Western imaginary of the period. As part of the source materials for an anthropology of adventure we also have to appreciate that, whatever the lack of literary merit, Norwood brings a particular set of concerns and desires to the imagination of the premodern.

Cold Wars and Hot Masculinities

It should already be evident that those concerns and desires were for Norwood, at least in *The Skull of Kanaima*, the presence of communist tyranny. Jacaré is in fact a Cold War warrior endowed with an instinctive sense of the rightness of such fine masculine figures as Rocks O'Neill and Jim Trent, and no less aware of the decadent evil of the unprepossessing Russian and German characters. As for many in Britain in the 1950s, the red-threat was compounded and confused by the very raw memories of the war against Nazi Germany that had ended only six years before Jacaré burst on the scene. As a result, it is not the contrast with brutish primates but with brutish communists and fascists that provides the canvas on which Jacaré's innate humanity, despite his lack of speech, can be drawn. Rather than a fear of closeness to apes, it is the possibility of slipping into communism that threatens the otherwise "decent" world of Norwood's Britain. By extension, the prospects for British and American development, as through the exploitation of the uranium deposits, might realize democratic order even in the "green hell" of Amazonia.

In the absence of the ape-society that was so crucial to the establishment of the Tarzan character, it is the insinuation of modernity and its communism that provides the mimetic justification for the violence of Jacaré. So recently valorized by world war, the fists and guns of Jim and Rocks likewise pummel and destroy the incipient new world order of Russian domination that is portended by a looming threat of nuclear aggression. At the margins of civilization this newly eternal struggle must be fought out, and the allegiance of Jacaré in that struggle is a way of naturalizing the rightness of western democracy. Its agents, in the form of Rocks, Jim, and Pamela, are thus able to disarm indigenous suspicion and fear, oppose the counter-discourse of communism amongst the still-heathen tribal peoples, and sustain the centrality of individual profit—since they are being paid for all this.

It is, however, a dangerous, if brave, new world and both Jim and Pamela's father are killed in the course of the story, so emotively justifying the otherwise questionable acts of violence in which Rocks and Jacaré indulge. But these are, in the words of the Jacaré film poster—"men of guts versus prowling death"—so that the intellectual issue of ideology is not at all part of their own motivations, which are rather more centered on the wholesome values of free enterprise and the redemptive power of just violence.

At the same time we cannot ignore the role of Victor Norwood's own peculiarities in the fashioning of the Jacaré series. For the hot masculinity that surges through the pages of these works was apparently a complete inversion of Norwood's own sexuality. I do not propose to attempt some kind of Freudian psycho-critique of Norwood's fictional writings,[7] but the prominence of masculinity in both his travel writings, especially the 1956 work *Man Alone,* and the works of fiction is, in this light, notable. In the travelogues of his own sojourns in British Guiana, Norwood certainly inhabits the world of Jacaré—everything is threatening, physically or morally, and only the masculine qualities of bluff and muscular action and decision-taking, enable Norwood to avoid "going bush." That such a masculinity was nonetheless problematic for Norwood is then suggested by his authorship of the erotic novels *Sex Gantlet (sic) to Murder* (1957) and *Hell's Wenches* (1963), since the sexualities in these works are entirely transgendered. Indeed the 1957 work was also released, under the pseudonym Mark Shane, as *The Lady was a Man*. In this work the repetitive sexual act is one of fellatio performed by cross-dressing men on other men so that one cannot help but wonder at the basis for this obsessive interest in a particular sexual act. The intellectual innocence of the childlike mute, Jacaré, thus contrasts with the bodily authority of the savage man that he also is. One might speculate that this interplay between a political innocence and masculine bodily rectitude, demonstrated in the measured forms of violence, was key in Norwood's literary muse, as others have argued for the pulp genre more widely (Parfrey 2003).

For these reasons, Jacaré is not really the kind of figure that Tarzan is in terms of how such fictions relate to the imaging of the modern and the primitive, or of the basis for humanity rather than animality. Instead Jacaré is more like an avenging spirit of the simple, innocent, and untutored in a world of foul complexity and transgressive identities and, as such, reflects a post-WWII condition rather than the post-Imperial context of the earlier works of Burroughs. The postcolonial world is already in the making and Tarzan has become faintly anachronistic. Jacaré, precisely because of an almost absolute absence of personality or apparent mental life can therefore become a disembodied, even evanescent, phantom of the forest, whose violence is unideological, albeit extremely vengeful.

In this context the invocation of the idea of kanaimà becomes somewhat more intelligible, although the forms and practices of this ritualized killing are not at all present in the content of the story itself. Rather, the allusion to kanaimà stands for the pure primitive and wild at the precise moment of its (presumed?) disappearance in the emerging postcolonial world. Jacaré is anachronistic because the moral force of violent hypermasculinity is insufficient to meet the threat of nuclear destruction, even if it forestalls it temporarily in the jungle margins of brave new worlds. The mismatch of muscles and knives against the unseen poisons of communism and radiation requires that "kanaimà" also becomes disembodied, a comic book horror with a fading reality as the rigors of modernity erase the basis of native autonomy.

This was also the way in which the idea of kanaimà was used as a literary trope by more credible literary figures, such as Rómulo Gallegos, the onetime president of Venezuela, in *Canaima* (1935, 1984), a novel set in the Caroni and Yuruari rivers during the 1930s. Apart from Gallegos's stature as a literary figure, the relevance of this work for understanding Jacaré is that in Gallegos's *Canaima* we find no longer the lone "Indian avenger" but the very spirit of the deep forest itself, not the elfin spirit of eco-fantasy, but a violent, masculinized were-jaguar.

Since Gallegos's *Canaima* inaugurated the metaphorical use of the indigenous notion of kanaimà as a literary trope of wildness, savagery, and supernatural malevolence, there have been increasing references to kanaimà as a backdrop for the imagination of the Guyana Highlands (Petit 1998), and as a literary device for suggesting the upwelling of atavistic and ancestral forces into the "modern" world, apparent in Wilson Harris's (1995) short story *Kanaimà*, first published in 1964. Less informed by the actualities of Guyana than Norwood, Gallegos, or Harris, but still entranced by the trope of "Kanaimà," Eugene Orlando (2000) offers a web-based collection of "Victorian romance" and adventure that includes one entitled *Kanaimà*. The tale is something of a "bodice-ripper" and so compares fruitfully with Norwood's Jacaré for its use of the notion of kanaimà to dramatize cultural opacity. These literary renderings are thus unwittingly embedded in old colonial tropes of kanaimà as indigenous, evil, miasmic, and ever-present—and in this way is actually a mimetic reading of Native thought itself.[8]

Norwood therefore participates in a wider discourse on kanaimà and one infers that his own experiences in British Guiana were the basis for this. There is no evidence in his writings that his experience of kanaimà was anything but literary or textual, however, the pervasiveness of this notion in Guyanese culture both then and now suggests it could hardly have been ignored. In this way the significance of the writings of Victor Norwood, and the Jacaré series, is intimately connected to the wider cultural milieu from which he emerged; the postcolonial world of uncertain sexualities, incipient tyrannies, and fading regimes of Western authority.

Conclusion

We should not imagine that, despite the execrable racial sentiments expressed and the bald plagiarism of others, Norwood was without popular significance. He was also the author of some twenty-six Western novels, in which "hot masculinity" was again a key motif. This quite impressive textual production, implying a continuing commercial viability to his works, was sustained over a period of two decades. Norwood, therefore, was notably more productive than some of his better remembered contemporaries (see Server 2002). In addition, he published with both British and American presses[9] suggesting that a transatlantic breakthrough was contemplated if not finally achieved. Perusal of some of the sources for the growing scholarship on pulp fiction and comic art suggests that he is also part of the emergent canon. Unfortunately Norwood's profligate use of pseudonyms[10] make assessment of his overall career tedious, and undoubtedly the quality, rather than quantity, of his writings has deterred most scholars up to this point. However as Haut (1995) puts it, "taking the genre too seriously can be as much of a literary crime as not taking it seriously enough" and in the genre of "hard-boiled" crime novels of the 1940s and 1950s Haut has argued that:

> Pulp fiction has always been concerned with the relationship between capital and crime, corruption and power. It can be construed as political but never proselytizing. Moreover it is a literature written by the proletariat for the proletariat. Several leading pulp novelists of the Post-War years were former employees in the aerospace and oil industries of Southern California and paid-up members of the Communist Party. Why else would the Committee of Un-American activities take such an interest in what was seen as lowbrow art form? Happily for the Committee, the authors may have had critical dissenting voices, but their characters were sick. In the hard-boiled crime novel of the Forties and Fifties, politics is invariably investigated by the alienated or the psychotic. Ideas were just as easily pulped as the characters. Hollywood had far more interest in such writers of disposable entertainment than McCarthy, in the end (Haut 1995: 17)

Importantly then, Norwood can be seen to have firmly placed himself on the opposing side—advocating the structures of power and criminalizing its opponents. Rocks O'Neill and Jim Trent are quite certainly not the psychotic dissenters of the crime novel, but rather the upright heroes of anticommunist struggle. This may also reflect the differences in pulp novel genres, since the jungle canvas is open for peopling, a relatively blank space on which any notion might be convincingly projected—as in the Jacaré novels. By contrast, the cityscape is already known and prefigured in readers' minds to a far greater degree, being inflected with each individual's experiences of the urban context.

Jacaré may not be an anthropologically sophisticated exemplar of the idea of natural man, but he evidently was a popular one. However, the notion of "adventure" may be more strongly present in Jacaré than even Tarzan, for the key difference here is the relatively large role given to Jim and Rocks as protagonists, whereas for Burroughs the Tarzan figure himself is always at the center of plot development. For Norwood, in fact and fiction, "adventure" is the perquisite of the colonial white man, not his jungle alter ego. In this way Jacaré is no more than a cipher for the intrusion of the primeval, and Norwood is apparently more intent on naturalizing his own obsessions than in depicting the hereditary superiority of even a wild white man, as in the case of Tarzan. This serves in turn to highlight the ethnic-ambiguity of Jacaré himself—his "Greekness" at least hints at preppy ancestry and his fair hirsuteness also signals a dash of the Aryan; but no tale of origins is given in *The Skull of Kanaima*—such mysteries must await a further reading of the canon.

Notes

1. The word *jacare* is derived from the Tupi language and means "alligator" in Brazilian Portuguese.

2. This was a common misapprehension about the nature of Amazonian trophy heads until Michael Harner made a close investigation of Jivaro head-shrinking, showing that the skull was actually removed before treatment.

3. Apparently a version of *Warawete*, the name of a Warao clan at the mouth of the Orinoco River.

4. Nor was Jacaré perhaps the most egregious case of plagiarism. For example, Marco Garron's *Azan the Ape Man: The Missing Safari* (London: Curtis Warren, 1950) barely even bothered to rename the Burroughs character. In the first of these adventures Azan was considerably closer to the original than Norwood's Jacaré. The oversized paperback's pictorial wrapper shows Azan confronting a fiercely fanged ape, itself a traditional jungle fiction motif used by H. Rider Haggard and, of course, Burroughs. Somewhat more outré is the fact that Burroughs's *Tarzan* series has recently been reissued in Australia by a publishing house called *Norwood*!

5. Frank Buck also made *Tiger Fangs* (1943), *Jungle Cavalcade* (1941), *Africa Screams* (1949), and *At War with the Army* (1950). *Jacaré* (1941) was billed as being "An adventure movie actually filmed on location in the Amazon jungle," and so somewhat contradicted the representations of Norwood himself.

6. Unfortunately a copy of this has proved impossible to find to date as it would have been relevant to consider how the location filming in South America played off the oddly Africanized text version.

7. Norwood was also a prolific author of Western novels under various pseudonyms such as "Jim Bowie" and "Vince Destry"—clearly plagiarism was a working methodology for Norwood! (see also note 10)

8. For a much fuller discussion of these issues see Whitehead 2002.

9. Norwood's *Night of the Black Horror,* a sci-fi novel, was translated into German the series *Utopia-Großband #198* under the title *Die Nacht des Grauens* (1963) übersetzt von Heinz Zwack.

10. Bibliographical sources (e.g., Clute and Nicholls 1993) list the following— Coy Banton, Sane V. Baxter, Jim Bowie, Clay Brand, Victor Brand, Ella Howard Bryan, Paul Clevinger, Walt Cody, Shayne Colter, Wes Corteen, Clint(on) Dangerfield, Paul Dangerfield, Johnny Dark, Vince Destry, Doone Fargo, Mark Fenton, Wade Fisher, G. Gearing-Thomas, Mark Hampton, Hank Jansen, Nat Karta, Whip McCord, Brett Rand, Brad Regan, Shane Russell, Rhondo Shane, Victor Shane, Jim Tressidy— and there is no reason to suppose at this point that such a list is complete.

CHAPTER 8

The Work of Environmentalism in an Age of Televisual Adventures

Luis A. Vivanco

"DANGER! DANGER! DANGER!"

        ("Crocodile Hunter" Steve Irwin)

"I'm Jane, he's Tarzan"

        (Terri Irwin, referring to herself and husband Steve Irwin; Simpson 2001)

"Somewhere along the way, there was an understanding reached between Tarzan and his followers. Either it was a collaboration—'don't bother me and I won't bother you'—or it was true conquering that was in some ways permanent. There must have been this understanding or there would not be so many Tarzans today."

        (Theroux 1998: 58)

Introduction

For millions of American television viewers, the dramatic expression "DAN-GER! DANGER! DANGER!" is identified with one person: the khaki-clad Crocodile Hunter, Steve Irwin. Irwin's Discovery Channel wildlife program *The Crocodile Hunter* has been rapidly growing in popularity (and notoriety) since the mid-1990s, with at least one estimate suggesting as many as two hundred million viewers in sixty countries worldwide (Simpson 2001).

The Crocodile Hunter phenomenon is based on the quirky Australian's ability to find, capture, manipulate, and explain dangerous reptiles and animals in exotic locales. He is not a scientist, nor does he claim to be one, but he is what one admiring fan I know calls "a naturalist on crack." His life seems a perpetual eco-adventure, roaming wilderness areas in search of whatever unique or threatened wildlife he can identify, grabbing and holding the world's most venomous snakes by their tails, and—his trademark activity—wrestling crocodiles, sometimes single-handedly, to ensure their safety, he claims, from poachers by placing them in zoos.[1]

To describe Irwin as one of his critics has—"It's the same old alligator-wrestling mumbo jumbo.... It has more to do with a carnival act than nature" (White 2001)—is to highlight the central place of nature-as-spectacle in Irwin's work and visual project. But this criticism sets up a false dichotomy between *The Crocodile Hunter* and other wildlife films and television, ignoring the fact that all nature documentaries are in some sense "docudramas" (Bousé 2000). It also simplifies the more complex intentions and meanings of what Irwin himself has described as his "documentary adventures" (Gora 2002), the nature conservation and educational aspects of this work, and the linkages drawn in it between adventure and environmental activism.

My central contention is that adventures and adventurers have long been lurking in environmental philosophy and activism, and Irwin's macho flamboyance makes for the latest and most visible manifestation of this long-standing tendency. But there is more to the character of the Crocodile Hunter than initially meets the eye, even if the visual medium emphasizes that meeting the eye is more or less all it should do. This chapter explores the complex relationship between what meets the eye, the assumptions and intentions that underlie the medium of film through which Irwin's conservation messages are created and distributed, and the implications of environmentalism-as-adventure for concrete ecopolitical action in the world. I scrutinize *The Crocodile Hunter,* noting the ways the visual medium's demands and forms of representation constitute and reshape the very nature of ecopolitics, emphasizing the fantasy spectacles of adventure over the hard work of collaborative social and political action in contexts of conflict and flux. My approach is multifaceted, and begins by positioning Irwin alongside familiar tropes and figures like Tarzan, "adventure," and "danger." I explore the relevance of these normative concepts in environmental activism, and then examine *The Crocodile Hunter*'s place in the history of documentary film's representations of nature and its problems. I conclude with an analysis of a particular film in which Irwin engages in a deliberate and specific kind of ecopolitics, a dramatic wildlife rescue in war-torn East Timor.

Tarzan and the Politics of Virtuous Adventures

Certainly for Steve Irwin, his sidekick/wife Terri, and their comrades, who typically remain in the background, living the life of the Crocodile Hunter is fear-inducing, adrenaline-pumping, and highly perilous work. But what distinguishes their work from purely self-referential pursuits of "adventure"—what one popular adventure writer describes as "the total reliance on oneself" (Stark 2001: 8)—is that theirs are "virtuous" adventures. They are virtuous both because of their self-conscious moral piety, and even more so in terms of the word's root meaning, of their potent virility. The first meaning is made manifest in the Irwins' frequent invocations of a higher calling, a morally righteous political mission on behalf of wildlife and natural habitat that unites Steve's unique ability to commune with animals with conservation advocacy and public education. His often-proclaimed goal is to raise awareness and knowledge about animals most people are usually hostile to in order to promote their conservation. In this sense, Irwin's work is reminiscent of previous adventurers—Carl Akeley, Martin and Osa Johnson, Marlin Perkins, etc.—who visualized and justified their "extreme" exploits in terms of their value to science, public education, and nature itself (Haraway 1989; Mitman 1999).

The other meaning of virtuous, as virile and potent, is one of the reasons that Irwin is often compared to Tarzan (White 1997; Simpson 2001), usually by his admiring fans. As Paul Theroux (1998: 48–49) noted, one reason that Tarzan is so compelling is that he combines a certain solidarity and sympathy for animals with masculine authority over them, making him a special, lordly, and powerful presence in the wild. Moreover:

> Tarzan knew about the jungle: each root, tree, animal, and flower, the composition of soil, the yank of the quicksand, the current of rivers. He had conquered by knowing and he knew because he lived in the jungle.... It was a kind of savage osmosis: he took the knowledge through his skin and he was able to absorb this wisdom because he was in Africa; he learned by being in the jungle (Theroux 1998: 49–52).

Irwin's life story similarly emphasizes his long-standing experience in the Australian bush and his extensive knowledge of dangerous reptiles acquired by handling them since he was a young child (Irwin and Irwin 1997). Yet, while Tarzan's element was the jungle, just as the Australian bush is Irwin's, neither is clearly *of* the jungle; in that sense, they are both outsiders, or as Theroux asserts, expatriates. For Theroux, what distinguishes Tarzan as an existential outsider—and Euro-American expatriates who are his latest incarnation in the modern Third World—is his delusion of superiority, racial and otherwise, and that others relate to him only to serve his own self-centered interests, uniqueness, and arrogance.

Theroux's critique emphasizes that Tarzan's—and the expatriates'— adventures are never morally neutral, naturally righteous, or even politically acceptable, and that they reflect tendencies of conquest over collaboration. The uncanny similarities between Theroux's Tarzan and Irwin's Crocodile Hunter—savage osmosis, virile authority over the wild, self-conscious outsider status—encourage one to consider how and why Tarzan-inspired adventure and its imagery lie at the heart of *The Crocodile Hunter's* virtuous environmentalism, and to ask what questions and insights these realizations might generate about the practical forms that environmentalism-as-adventure takes in the world. In Theroux's language, one can ask: what role do ideas about adventure play in the collaborations or conquests of environmental activism? What do followers of *The Crocodile Hunter* consider to be effective environmental activism?

Georg Simmel's seminal insight on adventure (1971), that it stands over and against the continuity of everyday life and thereby represents a particular social form in which fundamental categories of life are synthesized, is a significant aspect of the approach I take. Simmel provides a constructive means of considering basic assumptions, categories, and representational possibilities for human interactions with the natural world. The adventurer is defined by his or her engagement with the exotic and the unknown, and so the search for adventure can entail a deliberate and self-conscious attempt to reveal and revel in the Other—peoples, landscapes, states of mind, spiritualities, animals, and so on. Danger and risk are also typically at the center of adventure's distinction and significance, as are notions of deviance and liminality (see Gordon, this volume).

But what if we take seriously Lianos and Douglas's (2000) proposition that deviance can no longer be treated as marginalized behavior by marginalized persons, because no modern space exists without a consciousness of dangerousness? That is, as a sensibility for threat, risk, and danger has grown in late modernity, the line between risky, aberrant behaviors and normality has been blurred, if not erased. To be sure, the situations in which Irwin puts himself are audacious, but following Lianos and Douglas, his behavior cannot be simplistically written off as marginal. This is especially true because of the program's immense popularity and the fact that it is increasingly imitated and routinized by other television personalities, such as Jeff Corwin and Mark O'Shea, both of whom also appear on the U.S. cable television channel, *Animal Planet*. In a "dangerized world," where communal spaces (from the public park down the street to the Internet to the freeways on which people commute) are redefined as risky and fear-provoking, adventure—a self-conscious experiential engagement with danger and risk—represents an acceptable and increasingly significant, even ubiquitous, interpretive framework for social interaction. In this context *The Crocodile Hunter* serves a unique purpose, juxtaposing the ongoing redefinition of certain spaces and types of people as dangerous, with a more familiar, even nostalgic, version of danger—nature as a dangerous and exotic Other, as a site where humans can experience true, unmediated, and politically unproblematic adventure.

One of *The Crocodile Hunter*'s messages is that while these animals are dangerous to humans, humans are even more dangerous to them, because of anthropogenic habitat destruction and prejudicial attitudes. This process of "differential dangerization"—animals are dangerous, people are more dangerous—normalizes a familiar universalist and ahistorical vision of human-nature relations, of humanity inherently separate from, although with dominion over, wildlife and their habitats. As an "extreme adventurer," which Simmel defines as an ahistorical individual, a person living in a constant present, Irwin is a self-conscious outsider to nature, but less so than everybody else, who in his vision should look but not touch. In these ways, he advocates a vision of wildlife conservation that makes little distinction between humans and their histories, emphasizing their essential dangerousness for wildlife, while de-emphasizing the particularities and contours in how people think about and relate to wildlife and habitat. At the same time, I argue, he offers a depoliticized vision of the importance of expert mediation, made all the more natural by his self-portrayal as a nonexpert outsider to the scientific and conservation establishments. As Lianos and Douglas suggest, one aspect of dangerization is the promotion of a consumerist ethic, in which relationships built on trust and neighborly collaboration are increasingly deemed irrelevant, if not altogether eliminated, in favor of an emphasis on amoral market-mechanisms that promote safety. The dangerized and adventure-centered worldview of *The Crocodile Hunter* consistently invokes a message of safety, of the "don't do this at home" variety, reinforcing an armchair consumerist appreciation for wildlife. Even more importantly, its worldview and global reach reinforce an expatriate conservation, top-down and consumer-driven environmental activism based on a dismissal of—even conquest over—the complex local diversities and collaborations that are the precondition for any effective environmental initiative.

Environmentalism as Adventure

In an age in which adventure travel and support for nature conservation appear to combine in "eco-tourism" and magazines feature "extreme content" like *Outside* and *National Geographic Adventure*, it is productive to consider the ways in which environmentalism itself relies on and draws from the imaginations and practices of adventure. Eco-tourism, that niche of the global tourism industry that markets adventurous encounters with nature as its product, is increasingly presented by proponents as a privileged means to generate appreciation for conservation among the elite who are its consumers, as well as financial resources for the activists seeking to protect natural environments (Munt 1994; McLaren 2003). Eco-tourism draws on deep cultural and political currents in which adventure and adventurers have long been comfortably articulated in environmental philosophy and practice. Consider, for example, the establishment of game reserves in colonial-era East Africa. While these parks are today defined and celebrated as sites of wildlife protection, one of the main reasons they were established was to indulge bored colonial military and administrative officers itching for a big game hunt without competition from native peoples (Adams and McShane 1992; Guha 2000). Is there any significance to the fact that Russell Mittermeier, the president of Conservation International, an organization dedicated to the protection of endangered biodiversity hot spots in over thirty countries, has confessed to a lifelong Tarzan fixation, and that its vice-chair is the actor Harrison Ford, who played Indiana Jones (Choudry 2003)? What kind of articulations are there between environmental activism and adventure?

According to Simmel (1971: 191), adventure is defined by its capacity, in spite of being accidental and isolated, to have necessity and meaning. He advises, "Adventure, in its specific nature and charms, is a form of experiencing. [But] the content of the experience does not make the adventure" (1971: 194). In the realm of environmentalism and conservation advocacy, having an experience of adventure in the natural world is not by itself enough to make authoritative claims on behalf of nature, especially given that contemporary discourses on environmental degradation and problem-solving rely heavily on specialized scientific knowledge and claims-making (Hansen 1993). But the existential experience of adventure, especially in a wilderness setting, has been associated with environmentalist discourses and claims to speak and act on behalf of landscapes and animals in sociopolitical arenas, at least since the early twentieth century. In these discourses, nature represents an independent, even transcendental, place where the unexpected and uncontrollable occur outside of human will, provoking personally transformative experiences in which individuals can realize their spiritual unity with, and make a political commitment to, the wild.

There is an influential strain in Euro-American environmental thought, which includes writers and activists from John Muir and Aldo Leopold to latter-day deep ecologists and wilderness poets like Gary Snyder, that emphasizes that at the heart of modern environmental degradation and social anomie is the denial of the human connection to the wild. According to this perspective, a necessary precondition for the healing and redemption of both nature and humanity is opening oneself to the wild's transcendental powers through what we might think of as an experience of adventure. Following Simmel, it is not necessarily the content of the experience that matters, for its significance lies in the contrast it poses to the controlled and expected aspects of daily life. Aldo Leopold's (1949) classic and much imitated "conversion story" in *Sand County Almanac* provides a suggestive illustration. In it Leopold describes himself as a routinely unreflexive and callous hunter until the moment during a hunting trip when he is confronted by a profound wildness in the eyes of a mother wolf. In that accidental and unexpected encounter, he realizes the awesome independence, even defiance, of the wild. He informs readers that this encounter so unsettled him that it made him rethink his entire relation with the natural world, and he thereafter embarked on a career of fighting on behalf of the wild. Similarly, for Teddy Roosevelt, a powerful proponent of nature conservation (and hunting), wilderness provided a place for regeneration and renewal for men negatively affected by the physical and moral effeminacy of modern urban and industrialized lives (Mitman 1999: 13). There are significant practical and political distinctions between a Roosevelt and a Leopold, but they both identified the wild's transformative potential through the experience of adventure as a justification for the development and implementation of conservation interventions and managerialism.

The Crocodile Hunter's representations of wild animals similarly emphasize their existential independence and difference from the human world, while allowing for the adventuresome possibility of crossing the nature/culture boundary in order to appreciate and make claims on behalf of those animals. But Irwin's vision and ambition transcend any particular ecosystem or species, even nation-state, and his aspirations are internationalist. Indeed, with its rhetorical commitment to the cause of wildlife conservation around the world, *The Crocodile Hunter* draws on a long and familiar history of Western discourses of responsibility for global resources. In this case, it extends that responsibility to symbolically loaded and much-maligned wildlife like crocodiles and snakes. Julian Huxley reflects the internationalist vision and ideological justification in work like Irwin's: "Africa's wildlife belongs not merely to the local inhabitants, but to the world. To let it die or be destroyed would be to allow a precious element in that rich variety to be submerged forever in the drab monotonous flood of uniformity that is threatening to engulf our mass-produced technological civilization" (cited in Mitman 1999: 194).

Underlying this ideology is a shift in how Euro-American cultures interact with animals, from seeing them as labor or as productive raw materials, to animals as entertaining to look at (Berger 1980; Davis 1997; Mitman 1999); indeed this shift makes *The Crocodile Hunter* and other wildlife conservation efforts and film both possible and meaningful. As our guide into the dangerous and threatening zones of the natural world, Irwin is still an "outsider," certainly in relation to the animals he targets where he is a visiting, if dominating, adventurer, and even more so in relation to the people who sometimes appear at the margins of his programs. Given the show's global reach, Irwin inevitably comes into contact with a variety of cultures and peoples, who may have very different ideas than he does about the animals and landscapes where they live. When he does acknowledge human presence, his perspective is often a "misanthropic" one wherein humans are an inherently destructive species, while animals are essentially good and innocent (Franklin 1999: 3). But as I will discuss in more detail below, in Irwin's worldview some humans are more threatening than others, and his films draw on another deep wellspring of environmental thought, of ideological and iconic differentiations between peoples (Hollinshead 1999). This is often perceived as the modern and progressive rationalities of (typically white Euro-American) environmentalists butting up against the primitive and backwards rationalities of (often dark Third World) rural people. Especially in a context of rapid ecological deterioration around the world because of population growth and expanding capitalist economies, the question of who will take responsibility for those deteriorating resources is, of course, the keen one.

At the Limits of Documentary Credibility

One must emphasize the fact that both Tarzan's and Irwin's adventures are primarily meant to be seen on a movie or television screen. Our knowledge of their escapades is thoroughly intertwined with conventions of visuality and filmic representation that one cannot understand the place of Irwin in the history of environmentalism-as-adventure until the ways in which the visual medium itself constitutes and reshapes the relationship between environmentalism and adventure are considered. As a species of television documentary, *The Crocodile Hunter* relies on realist styles of representation—techniques of filming, editing, and narration that suggest a transparency of the medium—marking them as distinct from the Tarzan fictions of Johnny Weissmuller films or more recent Disney cartoons. Indeed, *The Crocodile Hunter's* credibility (in the sense of both believability *and* integrity) depends on these realist techniques; otherwise they can be dismissed as merely entertaining fictions. As Irwin's film director John Stainton says, "He's like a modern-day Tarzan. He really is. That's why people love the show—he's no glib actor. Steve's real" (White 1997).

But as Nichols (1994: 45) might say, *The Crocodile Hunter* is better understood as television that "aspires to reality," or as factual fictions where the pursuit of spectacle coexists with, and commonly outcompetes, objectivist narratives and realist imageries. Like other nature films, *The Crocodile Hunter* takes elements of the Griersonian documentary film tradition, specifically realist imagery and didactic discourse about the world in service to a mission for social change (Vivanco 2002), but breaks from its "discourses of sobriety" (Nichols 1994) with its style of high-energy entertainment and its pursuit of high status in the commercially competitive ratings game. It is close kin to the "reality television" genre although there are some distinctions as well. It is not of the same order as programs like *Survivor, Fear Factor, The Bachelor,* and their interminable spin-offs, which are organized around a game-show plotline and survival-of-the-fittest ideology. Rather, it is more like a combination of *Cops* and *When Animals Attack.* Like *Cops,* it proposes to teach viewers something about the world by situating the camera in a behind-the-scenes setting and taking spectators along for the ride with a friendly insider who explains to us what is going on. Like *When Animals Attack, Crocodile Hunter* trades on a widespread stereotype that the potential for wildness and brutality lurks just below the tranquil surface of ostensibly the most placid animals. The viewer's "reality" in *The Crocodile Hunter* is the vicarious participation in Irwin-cum-Tarzan's routines of eco-adventure and animal husbandry, and of alluring (yet comfortably distant) engagement with the explosive threat of wild animals portrayed in the program.

Following Nichols (1994: 54), Irwin's program creates a natural world of its own tele-visual devising, in which nature exists as a category and site of quintessential adventure, which keeps the realities and complexities of nature at bay. These lost complexities engender scientistic arguments that "nature is not like that," while other critiques point to the absence of any discussion of complex political and ecological contexts (as in the social, political, and economic pressures on wildlife).[2] Indeed, one of reality television's central goals is the production of a feeling of bewonderment, which simultaneously suspends disbelief and reality through the creation of a hyper-reality that visually exploits certain aspects of the world in service to a specific narrative. The result is that viewers can be involved in the morality plays associated with the televisual version of reality without necessarily confronting the gap between these narratives and the external realities upon which they feed, or the fact that for the televisual medium the only behavior that really matters is consumption (Nichols 1994).

The dilemma, according to critics of reality programming, is that social and political struggles in everyday life take their cues from the medium's constructions and rhythms (Miller 2000). As Nichols observes, "This ebb and flow of detached consumption, distracted viewing, and episodic amazement exists in time and space outside history, outside the realm in which physical, bodily engagement marks our existential commitment to a project and its realization" (Nichols 1994: 53). In other words, because the historical world is reduced to a set of simulations, the medium eliminates any historical consciousness eventualizing, because there is no logic or experience external to the program. Important as this critique is, the problem is that it is difficult to estimate how many people take their cues from *The Crocodile Hunter* in modeling their own practical involvements (or lack of involvement) with wildlife and its conservation. The program has reportedly stimulated travel to Irwin's zoological park in Queensland, Australia, although this does not necessarily suggest that there is a direct correlation between support for wildlife conservation and Irwin's films; only that some viewers are inspired to become tourists to see Irwin in person, or to otherwise participate in the global phenomenon of his celebrity.

In fact, in Irwin's case, viewers are attracted to watch him precisely because he is the one engaging in outrageous acts with dangerous and unpredictable wildlife, as a kind of proxy-adventurer who does precisely what "look-but-don't-touch" mainstream environmental discourse tells people they should not do. In my conversations with viewers of various ages (including preteens, teenagers, and adults) and a survey of various fan websites, I have been struck by the extent to which discussions of Irwin are consistently framed around his perceived madness, his accidents, and his outrageous stunts, not his virtuous conservation message. Rumors of Irwin's death are regularly discussed on fan websites (something that he even parodies in a Federal Express commercial where he is bitten by a snake and keels over in mock death because he did not use their rapid delivery service to send his antidote), suggesting a cult of the individual, not a cult organized around his ecopolitical messages. For Irwin, his perceived madness and the buzz it generates are justified because they provide a vehicle to communicate about the importance of wildlife conservation. But it is apparent that the entertaining spectacle overshadows the message of conservation, and instead becomes a vehicle for viewers to consume visual imagery of Irwin's ability to confront and manage potentially explosive danger.

The films knowingly create this effect, because the plot structures of *Crocodile Hunter* films tend to be primarily organized around the spectacle of Irwin's interactions with dangerous and unpredictable animals. In the process they actually marginalize particular conservation narratives, as well as the social and political dynamics that may inform us about the pressures on wildlife. Marchetti has argued that this process of marginalizing such narratives is typical of the action-adventure film genre that rely on the "myth of entertainment"—that these representations are meant to be fun and innocent—and not to provide serious discussion on social or political issues (1989: 196). She suggests that this triumph of action-adventure fantasy over reality is a reflection of ideological contradictions and tensions embedded in society itself: "Removed from the real relations of dominance within our society, fantasy can offer a 'safe' expression of ideological tensions so keenly felt by those outside of power and so often denied any real validity" (Marchetti 1989: 182). In this light, Irwin's appeal for many viewers reflects a response to the "look but don't touch" self-righteousness of environmentalism, a sense that he is transgressing boundaries many would themselves like to cross (but cannot, for legal or personal reasons). By focusing on Irwin's gestures and experiences of adventure—visually represented on film as a combination of luck and opportunism that Simmel long ago identified as a key characteristic of adventure—*The Crocodile Hunter* confirms that the act of consuming the spectacle of commercially motivated and routine adventures is more important than communicating about the difficulties, dilemmas, and opportunities of conservation. It is appealing precisely because it creates a fantasy world that seems so real.

Mission: East Timor

The contradictions and dilemmas of visualizing environmentalism-as-adventure emerge most clearly in a particular episode called *Crocodiles of the Revolution* (Stainton 2000). It provides a vivid example of how Irwin's virtuous adventures mediate between specific types of people and the animals in their midst; the intensely political, even militarily supported, nature of such work; and how his actions on behalf of wildlife are rendered into apolitical and ahistorical adventures by the visual imagery and program narrative. The episode is set in Dili, capital of East Timor, where military-paramilitary hostilities and guerilla warfare during the struggle for independence from Indonesia (1975–1999) have begun to subside because of the presence of a U.N. peacekeeping mission.[3] Under the auspices of the U.N. peacekeeping force, Irwin has come on a "wildlife rescue mission," a kind of international disaster relief and humanitarian mission on behalf of crocodiles, and by extension, the East Timorese people themselves. As he says toward the end of the film, "This is so good for the morale of the people; it's the first sign of the reconstruction of their city.... Mission accomplished, and what a fine mission!" (Stainton 2000).

The film begins with visual imagery of burned-out buildings and cars, familiar but largely nonspecific images that serve as a visual shorthand of a war-torn society for Northern audiences. Irwin narrates that he and several employees from his zoological park in Queensland have arrived there with a camera crew because the World Society for the Protection of Animals invited them to Dili to help rescue two captive crocodiles living in wretched conditions in the yard of a Catholic church. The situation of the crocodiles is pitiful—the water in the cramped pools is muddy and stagnant, and the crocodiles are malnourished, sick, and irritable—and so the unspoken but obvious message is that he is there to "rescue" the crocodiles from the Timorese themselves. Irwin (2001) explains that Australian soldiers ("Diggers") working as U.N. peacekeepers discovered the situation and then requested his help through the Society. The condition of the crocodiles apparently mirrors the condition of the Timorese people. Irwin narrates a brief history of the turmoil surrounding the East Timorese struggle for independence from Indonesia. The film's narrative avoids historical detail or depth, and there is little direct connection between the suffering of the crocodiles and the struggles of the East Timorese. But Irwin does mention Timorese cultural attitudes toward crocodiles. He says that the animals are considered sacred, chirping, "They love crocs! They believe their island is actually a solidified croc!" (Stainton 2000). Interestingly, he prefaces this statement with a suggestive comment: "The East Timorese people don't understand croc husbandry." His website elaborates,

> Both crocodiles were held in absolute [sic] appalling conditions, not even the worst dungeon would compete with the dreadful torture and terror they suffered. However, crocodiles have a sacred status with the Timorese—my dad and my aboriginal mates taught me that one should always endeavor to promote and enhance local peoples [sic] beliefs, traditions and sacred animals. Therefore, bringing the crocs back to Australia was not an option. I somehow had to maintain the Timorese passion, beliefs and sacredness, save these crocodiles and make sure they were maintained in a happy healthy state for the rest of their lives, which could be up to 80 years. Hmmmm!!! (Irwin 2001).

Irwin proudly reveals in the film that when he and his team showed up to begin its work at the churchyard, "Local people came in droves. They love their crocs!" (Stainton 2000). But what is even more remarkable is that, in spite of apparently high Timorese regard for crocodiles, Irwin, the filmmakers, and the U.N. peacekeeping forces deem them culturally ignorant of proper animal husbandry and conservation, and there are no efforts (on film at least) to consult with Timorese about the situation, much less discuss their attitudes toward crocodiles. In fact, the Timorese "droves" are passive and silent throughout the film, dark faces that are literally relegated to the background without opportunity of voice, commentary, or insight.

Timorese silence and passivity contrast with Irwin's activity: he plans and executes crocodile captures, builds new pens, and moves the crocodiles around the churchyard to place them in their new enclosures. Such action, in fact, takes up most of the film, and it has several implications for how we as viewers might come to know the problems of crocodiles and their protection in East Timor. First, one does not find out how Timorese themselves may have considered or practiced crocodile stewardship, much less what it means to say that they are "sacred." Second, the film also does not explain how a quarter-century of warfare throughout the countryside may have undermined the habitats crocodiles rely on to survive, or how the political instabilities have undermined the ability to provide adequate conditions for the already-captive crocodiles. This is because Irwin's reference to Timorese notions of crocodiles as sacred is an impressionistic and simplistic one, meant more to establish his credentials as a culturally sensitive operator than to consider how external circumstances (Indonesian and paramilitary aggression) might complicate Timorese stewardship efforts. Third, at the very least, one would assume that there is a complicated relationship between religiosity and attitudes toward crocodiles in East Timor, given that these crocodiles are being held at a Catholic church. The East Timorese Catholic Church has a reputation for its staunch pro-independence political stance, especially since the Indonesian invasion in the 1970s. That crocodiles may serve as the East Timorese Catholic Church's symbolic support for the independence movement and as conservatories of East Timorese cultural and natural heritage are themselves remarkable facts, and merit further attention if the goal is to understand what factors might impinge on crocodile stewardship.

Another remarkable aspect of the film is Irwin's reliance on the logistical and security apparatus of the Australian Defense Forces (ADF). As he says in the narration, "The crowd [of East Timorese] was huge—the Australian armed forces are there for security" (Stainton 2000), implying that perhaps the real problem is the East Timorese themselves, not the Indonesian-sponsored terror. At the same time, he expresses his admiration for the ADF "Diggers" and their righteous mission on behalf of the victimized East Timorese. These are shorthanded ways of acknowledging that his own adventure is made possible by the political-military might of the ADF and its role in the peacekeeping mission, and suggest complexities about the relation-ship between contemporary televised adventuring and the environmental and humanitarian efforts of U.N. peacekeeping missions.

Among those complexities are the ways in which the U.N. peacekeep-ing force operated in East Timor. Led by the Australian contingent, the force eventually numbered eleven thousand soldiers from various countries, who were there to restore and maintain security, to support and protect the U.N. authority, and to assist in humanitarian operations (Smith and Dee 2003: 45). While Irwin explains that Diggers invited him there out of "compas-sion" for the crocodiles, this is a gross simplification of who is allowed into an armed military intervention. Following its Kosovo peacekeeping mission, the U.N. encouraged greater emphasis on public relations and information sharing as a way to encourage greater international support for its peace-keeping missions (Smith and Dee 2003: 105). Although the public informa-tion campaign in East Timor was initially under-resourced and disorganized, Irwin's presence must certainly be understood in light of its relevance to the public relations mission of the peacekeeping force itself. Irwin's very pres-ence in East Timor, and the fact that he portrays his mission in the film's narrative as a humanitarian one of helping both crocodiles and Timorese reconstruction efforts, suggest a successful public relations campaign that was able to access new audiences ostensibly tuning in to a nature-adventure show.

It also indicates a suggestive confluence of international bureaucratic and military intervention and wildlife conservation, one that happens to fall within Irwin's own expressed strategy of conservation activism. As he explains in a *Scientific American* interview (Simpson 2001):

> [T]here is a new zoo strategy, which here in Australia we're taking very seriously. It's a regional approach, and Indonesia is part of our region. So any of the animals that are endangered or likely to become extinct be-cause of habitat destruction, we're pulling them into zoos—predomi-nantly rescuing the animals that are going to die anyway—and housing them, learning every single detail about how we can breed them and es-tablishing satellite colonies of that species so that we're ready when the cure *does* come, when we *can* rebuild habitat.

In light of this expansionist, regionally focused conservation activism, it is also instructive to note Irwin's rhetorical celebration of the Australian peace-keeping force, and to explore what their presence might reflect about the specific relationship between Australian regional geopolitical interests and the question of East Timorese independence. Irwin fails to mention long-term Australian geostrategic interests in East Timorese natural resources, much less its historically staunch official support for Indonesian annexation and its late interest in East Timorese independence. Since 1975, when Indonesia claimed East Timor as its own territory, Australian policy was to accept, and then in 1979 give *de jure* endorsement to, Indonesian annexation, making it the only Western state to do so. Initially, Canberra considered annexation a *fait accompli,* although economic and strategic prerogatives soon came to play an even more significant role, because of the development and exploitation of oil resources in the Timor Gap, which lies between East Timor and the northwest coast of Australia (Chalk 2001: 37). Diplomats from Jakarta understood Australian desires to extend its continental shelf to open up this exploration and development, and linked negotiations over East Timor to gas drilling rights for Australia. But with Suharto's forced resignation in 1998, and President J.B. Habibie's surprise announcement that he would be willing to let Timorese decide for themselves if they wanted autonomy within Indonesia (and his even more surprising announcement in January 1999 that if his offer of autonomy was rejected, East Timor could be independent), Australian policy began to shift toward support for self-determination and independence. This was helped along by various factors, including the Australian government's realization that East Timorese would not support anything but independence; Australian desire to take a leadership role in brokering a conflict that had been gaining increasingly intense international and domestic recognition; and growing revelations of pro-integrationist violence that exploded after the June 1999 self-determination referendum (Chalk 2001: 41–42).

Given that Irwin's own presence is situated within this broader history of official Australian indifference toward East Timorese self-determination, or that he relies on armed ADF peacekeepers to carry out his own strategy of wildlife conservation, it would be interesting, to say the least, to hear what East Timorese might have to say about him and his intervention to help the crocodiles. But not allowed any voice, much less an opportunity to reflect on what they think about the Australian zookeepers, U.N. peacekeeping forces, or the crocodiles in their midst, the Timorese occupy a position in which collaboration (one of Tarzan's options according to Theroux) is not an option, and instead are rendered passive actors in a top-down wildlife protection initiative—nay, spectacle—dominated by expert knowledge and technologies (not to mention the presence of armed Australian military personnel), with no regard for the political-economic inequalities and cultural realities of the historical moment in which they live. Unlike Tarzan, though, who generally sticks around, Irwin does not stay long but moves off to continue his adventures and apply his expertise elsewhere. In a very tangible sense, Irwin presents himself as the pinnacle of Western modernity, reflected in his commitment to an internationalist environmentalism whose knowledge and concerns are universal and ahistorical. As *The Crocodile Hunter,* Irwin's personal power and spectacle-inducing presence are such that the natives are silenced and even the Australian soldiers turn themselves into his helpful assistants.

Conclusion

In pointing out the nearly invisible political-economic and military structures that make possible Irwin's efforts in East Timor, we are perhaps expecting too much of *The Crocodile Hunter,* by holding it up to standards of reflexivity that are unreasonable given its medium. If these revelations are any indication of the conditions under which his other programs are produced, there is much more to consider about how his films render real-world events and histories into ahistorical spectacles of adventure, and by extension, what the meaning of "Danger! Danger! Danger!" might really be. But, to be fair, any analysis of the place *The Crocodile Hunter* occupies in contemporary cultural politics of wildlife conservation and environmental activism has to acknowledge the highly competitive commercial and institutional environments in which these films are created, and the projected audience sensibilities for which they are intended. In this environment, foundational concepts of human-nature separation and familiar narratives of human misanthropy and animal innocence inform the creation of images and plotlines. More importantly, these films draw on a deep historical tendency of seeing nature as a site of adventure, where the experience of the exotic, unexpected, and unknown are (ironically) routinized and expected, for the purposes of entertainment. Importantly, this reliance on adventure focuses attention on the vagaries of the moment, so neither history nor future exist in the visualization of adventure. This of course renders the spectacle as a natural and central point of the films, as we can see clearly in his film on East Timor. The point of that film is not to teach viewers about the dilemmas of conserving wildlife in sites of geopolitical flux and cultural difference, but to emphasize the personal heroism of Irwin-as-Tarzan.

Even though Irwin regularly reiterates his passion for wildlife conservation, *The Crocodile Hunter* films are unsurprisingly committed to keeping the dirty realities and difficulties of environmental activism and work in-the-world at arm's length, rarely offering any reflections on successful conservation techniques and projects. As viewers of *The Crocodile Hunter*, we are left with relatively few options. Animals are wild and potentially dangerous—and while a true adventurer would try to cross over into that world—most of us are content to let a proxy do it. And so we can rely on his expertise to tell us what and who is dangerous. Perhaps it is safest to let the Irwins of the world, using their technical abilities, handle the problem for us. Maybe the best we can do is continue to bear witness to his efforts on television. Or perhaps it is safest to visit our nearest Discovery Channel store and buy Crocodile Hunter souvenirs in hopes that he might dedicate some of the proceeds to helping wildlife around the world.

Notes

1. A longer version of this chapter was published in *Cultural Dynamics* 2004, 16(1): 93–115. Republished with permission.

2. This motivated one ecologist to recently coproduce what he saw as a "counter-documentary" for National Geographic's *Explorer* film series called *Quest for the Rainbow Serpent* (Wright 2002) about the Eastern Diamondback rattlesnake. The ecologist, Bruce Means, reports that after seeing Irwin's program he decided to make a "more sophisticated film showing the animals' biology ... and not jumping on them and doing all the other dumb things that are going on TV" (Ritchie 2002). He adds that Irwin and his imitators "don't do a lot to instill appreciation for animals. They prey on the fears and danger and overdo it for the sake of selling TV time" (ibid.). Means is correct—it *does* sell TV time to prey on people's fears and danger, as the boom in "reality" and adventure programming indicates—but what is important about Means's statement is that he seeks to create a distinct realm for his own film, emphasizing its virtuousness and representational objectivity. The statement elides, unconsciously or otherwise, the fact that all nature documentaries employ techniques and representational philosophies that create what one filmmaker calls a "period piece fantasy of nature" (Bousé 2000: 14; Vivanco 2002).

3. The U.N. military intervention was a response to a period of intense violence perpetrated on East Timorese by pro-Indonesian militias and Indonesian security forces, after a referendum in which the East Timorese voted against inte-

gration as Indonesia's "27th province," confirming a desire for independence and a public defeat of a quarter-century of Indonesian efforts to forcibly annex East Timor. During the most recent period of violence, dubbed *Operasi Sapu Jagad* (Operation Clean Sweep) by pro-Indonesian forces, an estimated three to four thousand East Timorese were killed, hundreds of thousands were displaced from their homes (including as many as 250,000 who were forcibly moved by the Indonesian military and integrationist militias to the Indonesian territory of West Timor), and as much as 70 percent of infrastructure was destroyed (Chalk 2001: 36; Smith and Dee 2003: 44).

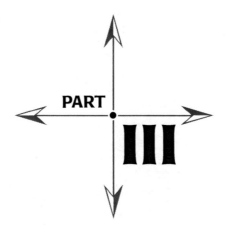

PART

III

High Adventures

CHAPTER 9

Five Miles Out:
Communion and Commodification
among the Mountaineers

David L.R. Houston

In the remote Western Karakoram mountain range is the plainly named peak of K2. It is the second-highest mountain on earth, with a reputation as a "killer" mountain. Climbing to an altitude of five miles is not a casual undertaking, but many attempt it. To reach K2 in 1953, climbers walked 150 miles through difficult terrain; today, they can fly almost to Base Camp. Today's expeditions are profoundly different from those of the 1950s, highlighting the contrast between yesterday's pilgrimage and purity versus today's commodification and status. This essay offers a perspective of mountaineering through the narratives heard and life lived with an adventurer, an experience that raises questions about why they climb and exposes differences between new and old-style mountaineering. Today, as an anthropologist, my investigation into this particular type of adventure is significantly challenged: how do I step back from the side effects of living in the shadow of the man and the mountain, escaping the emotional pull of family and self, to develop a critical eye toward the subject? In the larger sense, how do we as anthropologists escape the narcissism of minor differences, denying the similarities between ourselves and what we do and those that we identify as Other? The emotional relationship established between ethnographer and subject is multilayered, containing friendship and intimacy, aversion and hostility. It challenges our ethnographic sensibilities.

My exposure to mountaineering began at a young age. It may have been in conjunction with the release of my father's film about his 1953 K2 expedition, which ended in the death of a team member, Art Gilkey. This film must have been a significant emotional challenge for my father. The first scene is of my older brother—then a small boy—asking Bob Bates (another team member) about the trip; my older sister is already sitting nearby. I never appear in the film.

The film is the story of the expedition: how my mother made some of the gear; how the team—Tony Streather, Bates, my father, Dee Molenaar, Bob Craig, Pete Schoening, and George Bell—sat around our kitchen table carefully packaging supplies destined for twenty-eight thousand feet knowing that their lives depended on it. The film depicts my mother as a smiling, devoted wife, a former nurse with energy to help prepare for this, my father's last climbing expedition, but the first she directly participated in. Mountains and altitude played a dominant part in my family's life. My father's interest in mountains and altitude continues today as a renowned expert in the subject of high-altitude medicine.

Most of us have never climbed a high mountain, instead living in flat, even terrain. There is little need to climb mountains to gather food, have sex, or raise a family. Why do it at all? At some level, mountain climbing must comprise either a sanity test or an amazing adventure. Simmel suggests that the experience of "adventure" is as it is in part because it is defined in terms of its "difference in relation to our whole life" (1997: 220). Living a quiet life in ordinary places, a "step outside" poses a challenge to the rhythms of ordinary life. Most of us do not seek to deliberately upend our lives to risk injury or death. Routine is comfortable, and adventure, as Simmel notes, is "like an island in life" (ibid: 221), a place we do not visit often, something special, unique, perfect, or pure. Others return repeatedly, redefining or renewing that "difference" on each successive visit.

As a doctor, my father often faced a certain kind of risk and danger. His mountaineering, though, remains vague and distant for me, and I look back and wonder why he did it. If Simmel is even slightly on track, if adventure really is defined as that difference, then that defining difference was never present for him in medicine. Was there something altogether different going on? It was not a lack of exotic, faraway places to visit. He trekked to K2 in 1938. In 1950, he, my grandfather, Betsy Cowles, and H.W. Tilman reconnoitered the south side of Mount Everest. These activities provided a steady supply of peers and a camaraderie that my father obviously enjoyed. In 1953, he led the ill-fated K2 expedition. What was the pull? Why did he depart on an expedition of such scale barely one year after I was born?

Figure 9.1 • K2 expedition preparation.

Ideas about the "why" of mountaineering abound. David Breashears, Himalayan expedition veteran and the first American to summit Mount Everest twice, includes deprivation, challenge, risk, and what one learns about oneself on his list (Coburn 1997). There is an element of a quest, at times on par with a pilgrimage or religious experience (van Vuuren 2000: 9). George Mallory is better remembered because of his quip about why he climbed Mount Everest in the 1920s—"because it was there" (ibid: 9)—rather than that he attempted it all and died trying, his body turning up seventy-five years later in a macabre testimonial to the "sport." There are seemingly as many different reasons to climb as there are climbers: military campaigns (Houston 2002), ancient burials (Reinhard 1996; Reinhard 2001), and pilgrimages (Elizondo and Freyne 1996) as well as political control all make the list. The potential for problems and hardships makes it clear that it is not always a matter of necessity or practicality, the pragmatics of culture notwithstanding (Sahlins 1976). The personal nature of why mountaineers climb was a subtle backdrop in my father's circles. Conversations rarely spoke directly to "why." Instead, talk turned to the challenge or the distressing change in the spirit of climbing. There was never any single rationale for "why." One clear theme emerged over the years: the team. Changes in the spirit of climbing reflected their view of today's climber.

Devotion or Devotional?

To explore the underlying motivation for climbing as adventure suggests numerous possibilities and highlights similarities to other, decidedly unadventurous activities. Climbing entails an element of risk, possibly death. Expeditions demand effort and strong commitment, analogous to a religious pilgrimage. That mountaineering requires skill and training is evidenced in part by the numbers who are killed each year attempting it, and this in turn inculcates a ritual-like approach. Regardless of underlying personal motivation, irrespective of whether they climbed in 1950 or in 2000, climbers must inhere devotion, and to some extent, this has religious overtones of a devotional or spiritual nature.

Facing injury or death may be part of mountaineering's draw. Forcing the body to undergo physiological change far beyond normal limits entails measurable risk. The urge to climb must pull the climber past this if one measures success in summits. The documenting of routes up Mount Everest least likely to result in death (Huey and Salisbury Forthcoming) demonstrates this awareness. Despite the possibility of death as an excitement factor (Beedie 2003: 205), risk itself is complex: it is not just a matter of "facing down death." Risk acceptance is itself a process that varies with the individual as well as the activity, mediated by other factors such as money, advertising, status, and prestige (Castel 1991). Taking the risk to summit is part of the adventure factor in climbing, but risk alone does not fully explain motivation in climbing.

Looking past risk and danger, Simmel's idea of "dropping out of the continuity of life" (1997: 220) offers another perspective. Mountaineering clearly contrasts with the everyday, particularly in industrialized countries from which many climbers hail. The meaning of the climbing adventure may in part be understood because of the life led while *not* climbing. Lewis argues that "the meaning of one is defined through the other" (2000: 59), contrasting a "normal" life—work, family, community—with that of adventure. This contrast is itself variable in meaning and not simply a matter of "getting away from the office." It may offer an alternative with almost religious overtones, explicit for some (Messner 1980), less so for others (Krakauer 1997). Climbing is often more than making the summit; the entire process can be ritualistic. Turner's three stages of ritual (1969) are roughly synonymous with before, during, and after, a framework developed further by Beedie (2003: 209), reinforcing the idea of climbing as ritual.

As an experience that unfolds and draws structure through anti-structure, mountain adventure also has parallels to pilgrimage (van Vuuren 2000). The pilgrim leaves behind an ordered, complex system for one that is at once liminal and simple: "initiand and pilgrim cease to be members of a perduring *system* of social relations" (Turner 1979: 122, emphasis original). This underscores Simmel's "step out of the ordinary" for, as Turner notes: "In pilgrimage the pilgrim divests himself by personal choice of his structural encumbencies" (1979: 131). The pilgrim attains solidarity with place or person through the journey. This provides the distinct contrast to the ordinary and the ordered, and on return, the "consolidation" of the experience itself (Beedie, 2003). This requires time, and, much like the apprenticeship of the aspiring climber, long-term commitment. In the end, the reward is both the journey and that remembered, consolidated, and assimilated sense of *communitas,* the willingness of the pilgrim to accept the liminal state with the certain knowledge of an ending and transformation. For some, semireligious drive, a form of ritual or pilgrimage, or both may mark climbing.

Ajax Mountain in Aspen Colorado is a modest 11,000-foot peak. Developed as a ski resort in the 1950s, it was where our family lived for several years. Every Easter, we awoke at 4:00 a.m. and rode the Number 1 Chairlift to the summit. Sunrise was glorious, but our minimal religious ties gave way to a desire for hot cocoa before skiing down. For as much as the religious service mattered, there might just as well have been some sort of human sacrifice as a paean to religious authority. Perhaps my father did it out of obligation or community duty, but I now wonder if he simply loved the cold early morning ascent to a high summit. The insignificant role of religion in our lives leads me to wonder about the link between religion and mountaineering and whether or not it is religious for *all* climbers.

Despite our own religious vacuum, some elements of pilgrimage clearly resonate with mountaineering then and now. Post notes that "association with nature is a 'classic' pilgrimage theme" (1996: 4) and that living in harmony with nature is common in the motive of pilgrimage as the pilgrim tries to leave behind all of the disaffection with his or her usual life (ibid.). This separation and subsequent joining with nature—symbolic or actual— seems to comprise a core part of the rationale in climbing.

Pilgrimage frequently involves a period where participants are "betwixt and between" two different worlds or states. Death often figures in as the physical death of a key figure in the rite or ritual that evokes the pilgrimage, or a symbolic death of the pilgrim. The liminal nature of the process is conducive to the latter, and clearly, death is a possible outcome of mountaineering. That death can figure as part of the actual experience for the climber seems obvious, but we also can find parallels in the deaths of other climbers as part of the rationale and motive for subsequent attempts. Moving toward death, the climber must encounter something similar to any other dying process: an encounter with self that is transcendent. To have gone to that place and returned successfully is no minor event. Death becomes the ultimate adventure, replete with spiritual tinge.

Old versus New

Rereading *K2: The Savage Mountain* (Houston and Bates 1979), I recall visits from climbers through the years. Conversations center on wholeness, how the various members of a team helped one another, or on the power and awe of the mountain or journey. There is no talk of pilgrimage, no talk of religion or God, little mention of family. There is a low-level criticism about new climbers, ones that race up the mountain with the goal of summiting at any cost. This last aspect of the conversation is a common thread, emerging repeatedly, and more loudly, in later years. It is not surprising to find an implicit sense that the old ways are the best ways, that the intrusion of the new climbers is injurious to the sport as a whole as well as to individual pride.

There is a seeming divide between what I will call the "old school" and the "new school" of mountaineering. This divide appears, roughly, in the late 1950s and early 1960s (others posit different dates) and demarcates the team-based climbers from what I might loosely term the *scensum gratis gloria* climbers. This latter category deeply concerns my father and his peers. Others usually consider these "schools" as a single coherent collection. Colonial or imperial ideologies, militaristic or economic opportunism, and gender are all effective frameworks with which to understand mountain adventures (See esp. Ortner 1999; Lewis 2000; Bayers 2003). These motives and manners are undoubtedly central in the minds and actions of many mountaineers and, as ways of understanding mountaineering and its relation to others, work well. But as I listen to the voices from the 1950s and 1960s, it is striking how absent these particular frameworks are from the minds and emotions of these climbers. There is no trace of any recognition, appreciation, or even cynical opprobrium of such themes. There is an abundance of historical awareness; many of these same climbers served in World War II or Korea, in government, or as teachers. They are not staunch conservatives sweeping the marginalia of history away for the sake of national pride. Indeed, many are skeptics about the value and veracity of social institutions. There is a palpable disconnect for me here: totalizing schemes, despite the strength of their arguments, do not fit.

Beedie suggests that our habitus travels with us (2003: 206). What climbers live, they often take with them on their journey, just as pilgrims might. The kind of odd colonialism stemming from a life lived that seeks a departure from the ordinary is invariably part of the climbing adventure, existing within or in spite of any particular theoretical perspective. "Old school" seems to carry respect for both mountain and team, and embraces the departure from the comforts of the everyday and the escape from "civilized" life. There is a reverence for the majesty and remoteness, the setting and the wildness of it all; an almost holistic approach that itself becomes a practice (Snyder 1990). The "old school" defines the team-centered nature of the climbing through its collective experience, as one part of the entire process. There is a circumspect notion of pilgrimage, a recognition that one traces the footsteps of others who have gone before, that there is a change, a transformation, that takes place as a result of the effort, that the entirety of the journey is an essential part of the experience. "New school" mountaineering often entails expectation and manifestation of everyday luxuries (Beedie 2003: 206), reducing this same practice in what is a decidedly postmodern vein to a quest for individual glory as a component of, or mere exercise within, extant "civilization." The "new school" immerses itself in a new and different world. Beedie (ibid.) documents the use of guidebooks, maps, and brochures—a "tour literature"—that conveys a sense of anticipation and the "adventure-to-be" to the aspiring climber. In addition, he notes the profusion of technology and technological advances that are in part a valid (but not unexpected) extension of past techniques and an important selling point and marketing tool for today's climbing and adventure service provider (ibid.). The "old school" sees this as anathema, as trappings of civilization; this attitude appears in conversation and in experience. But this is not merely a simplistic matter of "old" versus "new." I would instead suggest that the old school and the new school invent and reinvent, and to a large extent *create each other*, and through the very process of being "at odds," and in the quest to best the other, end up instead lending mutual authority even through denial and disdain.

For "old school" climbers, there is a clear aversion to, even a rejection of, trappings that exemplify the "civilization" my father and his climbing peers sought to escape. This particular outlook is an integral part of the tension between old and new school mountaineering. Despite the irony, an "escape from civilization" is certainly part of how the old school defines itself, and the absence of same helps to define—at least in the eyes of many old school climbers—the new school. It is a clearly heard refrain in how the new school venerates the old, despite the embrace of the latest trappings and gadgets. This link to civilization is an important component in how the old and new coexist. Practice of mountain adventure today subsumes paid professionals in addition to sport enthusiasts. Martin Zabaleta, the first Basque to summit Mount Everest, related to me his despair over the changed nature of climbers and climbing, pointing out that the salaried professionals who take others to the summit today do so as a job, as part of a day's work. Behind this is a society that produces the technology, the machinery, the audience, and the means of getting there, expressly for mountaineering. Lest we forget the role of those that pay the bill, I would again cite Zabaleta: "if we take away [the guided climbers'] support, they can not [climb the mountain]" (Zabaleta, personal comment). Many of today's participants are products of industrial societies, and like the paid guides, venerate the "old school." Witness the fanfare and fervor that accompanied the discovery of George Mallory's frozen corpse on Mount Everest, despite routine encounters with other not so famous climbers' bodies. Paying a professional mountaineer to achieve the summit is an extension of the nonprofessional's regular—ostensibly civilized—life. It is not adventure in the sense that Simmel described it just because the would-be climber steps outside of the day-to-day regularity of a life. Instead, it becomes an extension of that same life, something one purchases or acquires, like an SUV or a cell phone, where the rationale varies. It is not an escape or departure from civilization, it is an extension of it, an accessory. An escape from death finds measure as bragging rights, another measure of success.

Escapism notwithstanding, climbers of any ilk rarely practice their profession in a vacuum. There is an unavoidable awareness of the past and the efforts of other climbers living and dead. The failures and successes are part of the heritage and history of mountaineering. This entwinement is puzzling in the context of how the "old" school and the "new" school see each other: there is an exchange taking place. Ortner suggests that the "dominant representations (and practices) have literally remade others in keeping with dominant projects" (1999: 19). While an important case can be made suggesting that mountaineers have, as the dominant party, remade others as part of their project (Sherpas are an example), this does not account for the relationship that quite clearly exists between "old" and "new" school. The question is simple: who is the dominant party? Do "old school" climbers hold sway over their younger counterparts and in so doing reserve the right of judgment and thus take a lead role in their (re)making or does the "new school" so venerate or disdain the elder group and so take that role? We can start to resolve this seemingly intractable problem if we consider that the two polarities both make and remake each other. The result is that *both* parties are dominant representatives; *both* are actively part of the process of making and remaking. The critique of the "new" by the "old" carries with it multiple meanings; we cannot simply dismiss it as the crank musings of an old, tired generation.

In some sense, the process of "getting away from it all" is a kind of ritual behavior. In industrialized society, earning a living can itself be a demanding (sometimes demeaning) and grindingly regular aspect of existence. "Getting away from it all" takes on a more urgent meaning, acquiring a status of its own based in part on *where* we go and the perceived status of the destination. That the industrialized countries operate as a system whose jobs and careers dole out chunks of time for vacation drives this sense of urgency, compartmentalizing the "getting away" process in a way that makes it part of the everyday despite potentially exotic destinations. This rote vacation then becomes a kind of ritual: get the plane tickets, take the cruise, climb the mountain. The entirety of preparing for, going to, experiencing, and returning from the chosen destination has ritualistic overtones. All of this is "instant," part of the pace of the same industrialized existence that demands the getting away. By contrast, "old school" climbers felt they had to commit patiently to the activity over a long term through apprenticeship (Beedie 2003). The aspiring climber was expected to serve in an ancillary capacity on early climbs. Today's climbing adventure is no longer something that the amateur participant can afford to commit to over the long haul. As with many other forms of "getting away"—exotic cruises, swimming with whales, trekking to remote locales, caravanning across Asia—mountain climbing is but one choice among a veritable cafeteria of "adventures," each one tempting the alienated worker-as-consumer with a special, personalized escape route from the uncertainty and mundanity of a workaday world.

Even as his age claims him, my father continues to write and publish work on high altitude. At ninety-one, with help, he completed the fifth edition of *Going Higher* (Houston 1998). He continues to meet with friends and peers. At a recent gathering, my father honored his friend Thomas Hornbein, first up the West Ridge of Everest in 1963. The guests included a mix of "old" and "new" school climbers. Exchanges were laden with little barbs and one-upsmanships, reflecting the gap between two vastly different philosophies. There is the bizarre story of the competition to see who could have sex at the highest altitude—a French couple is apparently the winner at just over twenty-six thousand feet. There is talk about the effort to build a "cyber café" at twenty-five thousand feet, where climbers can check email or get a cup of coffee. There is the tale of the Finn who decided to bicycle from Finland to Mount Everest, climb it, and then bicycle home. A little later in the evening, Geoff Tabin arrives and quickly commands the crowd's attention. He was the first Jew to summit Mount Everest in 1992 (Ortner 1999: 289); his climbing politics contrast sharply with Tom Hornbein's, distinguishing "old" and "new." The evening is a study in differences, a moment out of time where the shadow of K2 is suspended and the philosophical outlook on the crazy business of climbing is brought into sharp focus as a *lived* experience of those present.

Despite appearances, the old and new schools need each other. Without the "old school," the "new school" has little on which to base their comparisons and claims to fame. Does it make any difference if you are the first of a particular ethnicity to summit Mount Everest if the audience does not care, even if the response is in part a measure of difference? If there is no "young buck" to try and best the old-timers, does it matter at all if there was a George Mallory some seventy-five years prior? Climbing today is often a radical departure from how it was in the time of my father or Tom Hornbein. Technology and tourism have spawned not just a different way of doing it, but also a popular culture and literature that at once venerate and taunt the "old" and pulls the chain of the "new" as they demand a peculiar respect for the past. Contradictions abound: *Outside* magazine reports that after hosting some sixteen hundred climbers, Mount Everest is now characterized as a mountain for "sissies" (Fedarko 2003). In climbing as adventure, a brush with death is now expected.

Commodification

Despite their mutual need, important differences separate the old and new schools, reaching past the odd symbiosis, pilgrimage, or ritual. The theme running through shared stories among old-school mountaineers centers on communion. The *whole* act of climbing from start to finish emerges as a kind of communion with nature and with self, something outside the purview of "normal" existence. Ritual and pilgrimage make sense here. In both, the actor undergoes a transformation. In the process, the differentiation between ordinary and extraordinary is elevated, recast as sacred and profane. Durkheim observed that "anything can be sacred" (1965: 52). The mountaineer passes from one place to another. In the old school, this passage drew a sharp distinction between sacred and profane; the mountain and the climb itself became sacred, the climb *pure*. The climber, through a series of stages—apprenticeship, summit, return—was initiated. The visit to the mountain became a sacred act. As a contrast between old and new, purity figures prominently into the minds of the old school. The new school, frequently circumventing the expected steps required to attain this same purity, is almost held in contempt by the old school. There is a kind of violation of the sacred, a pollution of the pure. The purity found by the old school was in part a result of the almost secret nature of the places they visited, that those places were far away from the reach of the civilization they were trying to escape. This secretiveness is also linked to and an integral part of what I might call a "suspicion of the present": that the new school does not seek out the remote secret places because they are remote and secret, but rather that these places are no longer seen *as* secret places. In this commodification, the "new school" has violated a purity rule. The old school thus constructs a worldview of the present that finds civilization and, by association, globalization and late capitalism, as anathema, and, in turn, as impure. It is not simply a clash of values about mountaineering, but a fundamental difference over the nature of the world then and now, with the past as trump.

Fueling this gap is the manner in which climbing has itself been transformed. Mountaineering as adventure is now "technologized," just as skydiving and heli-skiing are. These activities are marketed as a type of getaway (Beedie and Hudson 2003). Such leisure activity creates a corresponding "market" for the destination proper, separate from the activity and all it entails. The idea that mountains themselves and mountaineering as an activity are commodities is not new (Johnson and Edwards 1994). The cost of today's mountain adventure makes this commodification obvious. It costs $70,000 just to enter the country to go to Mount Everest—exclusive of all other costs required to do the actual climb (Beedie and Hudson 2003: 629). Bayers (2003) suggests that one needs about $65,000 to actually climb Mount Everest. Demand for "highbrow" adventure reflects this increasing commodification. It was clear, at least in the opinion of the old-school members I listened to over the years, that this was an unwelcome trend. Commodification separates mountaineering from any sacred or pure nature it may be imbued with, at least in the minds of many old-school climbers. Those that buy their way to the top are, by implication, "impure" and unworthy. If the means by which climbers undertake the summit is somehow "corrupt," then meaning is lost.

Communion

Almost fifty years after his K2 attempt, my father and I were driving home from a long trip. He related to me how, one year after K2, he had been driving home on an unfamiliar route. He began to see streams of blood on the road ahead of him, and soon drove into a different town altogether. He stopped and walked aimlessly about, having no idea who he was or where he was. A police officer later recognized him, finally connecting to someone who took him home.

My father was deeply shaken by this episode, not so much by the loss of identity and place, but by the hallucination of the bloody road. "It was Art Gilkey's blood," he told me. For my father, climbing mountains was a special experience. In spite of his failure, he kept this experience with him, not simply as a remembrance of the past, but as a lived experience, one that he later remapped in many different ways. For him—and as he extended this to us—this was a form of communion with nature, something done with reverence and respect, not to be taken lightly. Despite our irreligiousness, adventure into those wild places was as close to any religion as I can recall. It was a kind of communion. My father's attitude toward the new climbers was respectful on the surface. Only when he was with his contemporaries did the other side of this emerge: the "new school" crossed a sacred line. As the uninitiated, their claims about their strength as climbers were suspect. My father's legacy is not one of famous summits. He continues his work in high-altitude medicine, studying sicknesses that kill climbers like Arthur Gilkey. This is, perhaps, as much an act of atonement as it is a love of medicine and research. It is an odd discovery I make in the appendix of the twenty-fifth anniversary edition of *K2: The Savage Mountain* that my father's sense of responsibility for what happened to Art Gilkey at twenty-eight thousand feet runs so deeply: the reunion trip made for the book reprint helps him get a perspective on this failure. This feeling never really leaves him. Death is the five-mile shadow cast over all of us.

Mountaineering has changed since 1953. Climbing today is commodified, an almost unavoidable result of the confluence of postmodern life and late-capitalism. It is difficult to imagine it being anything else. For large-scale expeditions, team solidarity and communion with nature through respectful travel into wild places, at least as it may have existed for the old school, seem like shadows of the past. Death is no longer something to avoid, it is something to be encountered and conquered. It is easy to critique the climber of today as little more than a pumped-up tourist on steroids, someone whose place in the world gives them a disposable income that can entertain a trip to Mount Everest as a means of "getting away from it all." The legacy of climbing is more than this. Many unsung, unknown, and uncelebrated climbers venture out to the dwindling number of wild places for the sheer joy of a self-designed, self-propelled communion, a pilgrimage all their own, circumscribed by ritual. They do not measure their success in status, headlines, or best-seller books, but in their hearts and minds, leaving their own legacies for their children.

Anthropologists set out to study from a distance those whom we are not. We enrich our practice through participation and observation, stopping just short of going native. The apparent dichotomy challenges us, and as Narayan notes, we are constantly "shifting identifications" (1993: 671). Often this shift is minimal: we maintain our "distance" and our claim to an anthropological core. At the margins of experience, however, this equation is more difficult. Here, we define the adventurer as "Other." In a peculiar, even perverse, way, the anthropologist is joined at the hip to the subject of study while simultaneously enforcing a distance born of professional codes. Is what we do so different from the adventurer? Our own identities *are* "multiplex" (ibid.: 673) and we are seemingly in denial. Adventure presents us with a challenge to our own denied "nativeness." Do we define ourselves exclusively in terms of how we distance ourselves from the Other we seek to study only in a context of a sterile pursuit of knowledge? If we deny a part of us that wears adventure like a well-fitted suit, we risk becoming the object of our study, our own "other."

For over eighty years a wooden plaque has hung over the mantelpiece of my grandfather's Adirondack camp with the first line from Bryant's *Thanatopsis:* "To him who in the love of nature holds communion with her visible forms, she speaks a various language." On a personal level, this quote speaks of the person my father was in the context of his mountaineering experience. An ethnographic account of adventure is a challenge to produce. Running contrary to the usual practice of participant observation, an account of adventure inevitably must be done from a distance. For me, this has left many unanswered questions, and a distance between who I understand my father to be and who I am through him. My own perspective of this activity is peripheral in the sense that I was never at twenty-eight thousand feet—I stood at a distance with no choice but to live this adventure not on the slopes of K2, but in the five-mile-high shadow it casts.

CHAPTER 10

Crampons and Cook Pots: The Democratization and Feminizations of Adventure on Aconcagua

Joy Logan

Normal life can be such drudgery; little seems important. But in the mountains, all is different, for the alpine life is a life of consequence.

Gregory Couch

Adventure has the gesture of the conqueror.

Georg Simmel

"*Aconcagua-toda adrenalina*" [Aconcagua: total adrenaline] was the phrase that stopped me dead in my tracks in the middle of a souvenir ship in Mendoza, Argentina.[1] How, I wondered, could this T-shirt slogan really be referring to the experience of mountaineering on Aconcagua, at 6,962 meters the highest mountain in the Americas? At altitude a mountaineer's pace is determined, steady, and most often laboriously slow over prolonged periods of time, which belies the heart-stopping kind of exhilaration that the T-shirt suggested. The only way to reconcile these two contradictory images was by imagining that the shirt was marketing Aconcagua to an entrepreneurial, as well as a sports-minded, audience. It seemed much more reasonable to me to attribute Aconcagua's adrenaline factor to the rapid and explosive growth of *andinismo* in the Aconcagua Provincial Park and to credit the heady experience of an adrenaline rush, not just to high-altitude mountaineers, but to those flatlanders leading the development of the Mendocino adventure tourism industry.

Yet understanding the adventure of Aconcagua in that way only led me to other kinds of questions: For whom had the T-shirt, and by extension mountain adventure, been custom fitted? Could any national subject don its form and message? Was it exclusively designed for a masculine subject, or was it also tailored to fit feminine subjectivities? In other words, to whom is mountain adventure directed? And how do women experience this adventure on Aconcagua?

These musings about the T-shirt slogan reminded me that there is no single understanding of the adventure-ness of Aconcagua, a seasonal global village where for four months of the year climbers from over sixty countries, as well as local professionals, come together to experience both the ordinary and the remarkable. In fact, it is that sort of reading of the multiplicity, mutability, and the capacity for cross-cultural adaptation of adventure that guides my study of the evolution of adventure and mountaineering in the central Andes of Argentina.

Tourism and Mendoza

From its inception as an elitist, First World enterprise, mountaineering has become a profitable option for adventure tourism in Mendoza, a predominantly agricultural and wine-producing region of central west Argentina. Both local media and provincial government policies regularly cite tourism, especially in terms of the international renown and popularity of Aconcagua, as a source of significant opportunities for the province to advance regionally beneficial international connections (Turplan 2000). During the last twenty years Mendoza's provincial government has invested heavily in an official mountaineering infrastructure to encourage the Aconcagua industry. The Aconcagua Provincial Park was formed in 1983 and subsequently educational, safety, and regulatory institutions and practices developed to accommodate the increasing climbing activity there. The Valentín Ugarte Escuela Mendocina de Guiás de Alta Montaña y Trekking (Mendoza School for Mountain Guides) was established, the corps of provincial park rangers, and the provincial high-altitude rescue squad, and statutes for regulating service providers and providing medical service were also formed and put into effect. Thus, the growth of mountaineering on Aconcagua that developed out of the mountain's reputation as the highest mountain in the Americas, a most uncommon, challenging and elite destination, necessitated Mendoza's efforts to modernize, regulate, and control the park and mountaineering practices.

Accompanying the economic and governmental promotion of Aconcagua has been a seemingly concomitant sociocultural embracing of the Mendocino mountain that goes against the grain of Mendoza history, custom, and practice. Traditionally, Mendocinos hold that Mendoza "turned its back on the mountain" to look away from the Andes and toward the east, agriculture, European immigration, and Buenos Aires for its future. Major tourist attractions for the province have heretofore been outgrowths of this tradition, as witnessed by the weeklong harvest celebration, *la Vendimia* [Grape Harvest] (held in March) and by the innumerable guided visits to local wineries (which number close to one thousand in the area). It is this sedate and agricultural Mendoza of tourism and history that contrasts with the local media's positive and extensive coverage of Aconcagua adventure. It is also this traditional understanding of Mendoza and its legacy as the *"tierra del sol y buen vino"* [the land of sun and good wine] that explains the magnitude of the statement in January 2003 by Fernando Grajales, one of the founders of the Mendoza climbing industry, that *finally*, after more than a century of mountaineering on Aconcagua, Mendoza now claims the peak for its own and subsequently has integrated a sense of the mountain into its culture (Garcia 2003b).

It is this particular statement by Grajales that motivates my consideration of how adventure articulates national, regional, and gendered identities. Ultimately, Grajales's sentiments provoke the question of what place adventure and mountaineering hold in the Mendocino imaginary.

Mountaineering Preliminaries

Some of the principal assumptions that popular culture holds about the adventure of mountaineering are succinctly expressed in the 2001 film *Vertical Limit*. In the film, Elliot Vaughn, the brash and aggressive American multimillionaire who has financed the expedition up K-2 for publicity purposes, admonishes his guide for suggesting their immediate descent due to an approaching storm. "What did you think?" he asks sarcastically, "that the mountain would just lift up her skirts for you?" This scene, in which Vaughn also uses economic threats to goad his guide into taking an unnecessary risk that subsequently leads to disaster, most explicitly emanates from an image of mountaineering where the mountain, a decidedly phallic symbol, is feminized, its domination eroticized, and the homo/hetero erotic tension of mountain climbing is played out as ritualized competition for masculine supremacy. At the same time, this scene also posits three other fundamental questions at the heart of the sport: is capitalist materialism the nemesis of true mountaineering, where and how does the Third World figure in this predominantly First World endeavor, and how do women participate in this duel of masculine competition?

This scene is indeed a Hollywood reenactment of what Sherry Ortner (1999: 11–36) finds to be primary characteristics of the mountaineering tradition: it generally occurs at great distances from Euro-Anglo American centers that understand these far-flung locales as significantly different in terms of "race," culture, religion, and "modernness"; the dominant discourse of the sport links its adventure-ness to a critique of modernity; and its practice has been an essentially twentieth-century game of defining the masculine self.

Aconcagua Mountaineering History

Mountaineering's conjoining of adventure-ness to questions of modernity and manliness began on Aconcagua in the late nineteenth and early twentieth centuries as part of the European empire-building moment. Narratives from this period that penned heroic masculinity into mountaineering adventure also shaped and justified the ideological underpinnings of imperialism. Peter Bayers claims that tales of early mountaineering exploits extolled masculine virtues such as bodily virility, rationality, leadership, self-sacrifice, ruggedness, and resourcefulness as idealized traits of U.S. and British manhood that justified Western expansion and superiority and defended national masculinities against threats of "over-civilization," urbanization, and feminization (2003: 3–4). It is out of this epoch and tradition that the exploration of Aconcagua, paralleling the incipient development of mountaineering elsewhere in the globe, was initiated by European and Anglo-American explorers, adventurers, and scientists.[2]

The long history of international scientific exploration and national expeditionary teams left its indelible mark on Aconcagua in their naming of routes, refuges, geological formations and camps. The Polish Glacier, Plaza Francia, Camp Canada, Camp Berlin, Conway's Rocks, Messner's Ridge, the French Route, and the Slovenian Route, among others, mimic gestures of empire as they commemorate the supremacy of the individual and/or the national groups who christened them. These acts of naming and claiming, these symbolic performances of colonization, domination, and masculine heroic adventure, represent a moment of intrinsic self-satisfaction for the adventurer engaged in what Bayers (2003) calls "an imperialist masculine ethos." At the same time, the principal extrinsic reward for their great expenditures—monetary, physical and of the spirit—was subsequent public acknowledgement of masculine and national superiority on the mountain in the name of adventure, in venues like the Royal Geographic Society or in the publication of an expeditionary narrative.

As facsimiles of conquest, mountaineering adventures on Aconcagua have been inscribed within the tradition of heroic masculine travel narratives, guides, and histories, and on the mountain proper as the Western experience of white male foreigners exposing themselves to great risks for personal satisfaction and, often, Euro/Anglo-American national pride. For them, adventure on Aconcagua was both extraordinary and unrestricted, not in terms of monetary gain, nationality, or gender, but rather in terms of distance from the everyday and in possibilities of transcendence. For those European and Anglo-American explorers Aconcagua offered that "island in life," as Simmel would have it (1971: 189), in which life itself is comprehended. Thus, early mountaineers sought out Aconcagua as a proving ground for the individual where challenges to the solitary self could be, and often were, elided to issues of heroic masculinity and national identity construction.

The Adventure of Nation and Gender

Today it is still this fusion of self and national discovery as by-product of mountaineering that functions as a master trope of contemporary Aconcagua climbing history as it is represented in the informal climbers' museum in the Plaza de Mulas "Hotel" Refugio. At 4,370 meters, the Refugio at Plaza de Mulas, the base camp for the most popular approach to the summit, known as the normal or northern route, climbers memorialize their experience on the mountain by leaving behind articles of clothing, national and regional flags, group pennants, pictures, posters, and personal art. The mountaineers choose their mementos and area of display independently and write their own narratives of dedication or memory to inscribe themselves into Aconcagua's history. The mountain artifacts are put up by the mountaineers spontaneously throughout the Refugio on lines crisscrossing the rooms or on walls, doors, and door frames. The main dining hall functions as primary site of exhibit and the front entrance, passageway, activity room, and climbers' kitchen at the end of the passageway as secondary sites. The majority of clothing articles are inexpensive T-shirts that bear personal messages of effort, sacrifice, jubilation, congratulatory rhetoric, and dedications ranging from father to son, climber to nation, and climber to future climbers. Flags, emblems, banners, and posters are also signed in this same kind of laudatory register and articulate their mountain experience as adventure in the manner that Swarbrooke (2003: 16) defines it: a heightened experience that induces intense emotions, that involves taking risks on an intellectual, physical, and emotional level, and that is intrinsically rewarding.

As reterritorialized space, Aconcagua is often represented here as a repository of inflammatory national and regional political commentaries. In 2002–2003 an especially vehement Basque nationalist inscription on a Basque separatist banner in the front hall passageway caused several vociferous complaints from other Spanish nationals who requested that the manager of the Refugio, and as such, ad hoc curator of the museum, remove it from view. The manager refused by stating that he believed in freedom of expression and felt that mountaineers had the right to leave on display whatever statement they wanted to make about their time and presence on Aconcagua. That the Aconcagua mountaineering experience would elicit this kind of political antagonism stems, on the one hand, from European modernist traditions that privilege distance and displacement for realizing notions of home, national reality, authenticity, insight and creativity (Kaplan 1998: 27–64). On the other, it presupposes and maintains the First World colonial legacy of mountaineering on Aconcagua. In general, the imperialist tone of this nationalist dialogue keeps with most international climbing representations of Aconcagua that merely function as a forum for individual and expeditionary self reflection and for national, and especially North-South, identity politics to be played out.

Tellingly, those identity politics are an almost exclusively masculine game as feminine subjects generally are absent or hidden in the Hotel Refugio's displays. The predominant article of clothing that is left behind by climbers as remembrance and documentation of their time on the mountain is a T-shirt. These shirts of all colors and sizes are nominally marked for gender as masculine or, at the very most, could be argued as unisex garments. Although women often sign the T-shirts, flags, and pennants left by expeditions in which they have participated, there usually are no overtly feminine items of apparel on display. However, explicitly male-gendered items are exhibited, such as the set of Stars and Stripes boxer shorts hanging in the front and center of the dining hall. Some of the overtly masculine displays are even quite aggressive. A pair of Luxembourg jockey shorts hangs in a prominent place on the wall beside the main door to the dining area. They are stuffed strategically in the front to insinuate male genitalia and bear the slogan "une raideur de avance" "a rising rigidity," which plays off, and one-ups, a nearby poster's proclamation about the "rigors of ascent." This evocation of male bodies and the overt corporeal/sexual competitiveness of the displays are directed to the predominantly male public of the mountaineering community at Plaza de Mulas. While the visual message of the displays connects to the tradition that Bayers finds in early mountaineering narratives where the construction of masculinity can be understood by the "elevation" of the supposed virility of the imperial male body to dominate the natural environment (2003: 5), it also connects to later mountaineering attitudes that, according to Ortner, consisted of a "heightened sexualization of the mountaineering enterprise (and of mountaineers as men)" (1999: 162). The displays create a spectacle of man-as-body where forcefulness, strength, size, and mastery set a standard for manliness. Ultimately, the displays produce a libidinous system of homoerotic tension in which the male gaze is directed to the mountaineering male body in a masculine game of desire as domination.

The explicit depictions of competing masculine bodies may have provoked some jocular commentary, but they have not sparked any public outcry of complaint. There has been no ongoing argument or criticism caused by these signs of masculinity as regional and national symbols have. Therefore, the Refugio's ad hoc curator has not felt compelled to censor or take down suggestive masculine articles of clothing, just as he has not removed inflammatory political statements. However, the only overtly feminine-marked garments that once hung in the dining area museum (a red panty and bra set) were removed by the Refugio's manager after a relatively short tenure there.[3] The manager states that the mountaineers made such a fuss over the exhibit by constantly removing the articles to throw them around the dining hall and using them as the butt of lewd jokes and raunchy comments, that he felt obliged to remove them to keep the peace (or, to reestablish the homoerotic order of mountaineering). Thus, while the display of international regional-national conflicts and politics and the playing out of competitive masculinities are allowed and highlighted in the Refugio's museum, the overtly feminine is seen as disruptive and disorderly, even subversive and dangerous.

This reading of the feminine in the Refugio's museum can be taken as an overview of how gender dynamics in Mendoza's mountaineering enterprise on Aconcagua work in general. The subversive nature of women's presence is stabilized by rendering them invisible, either by denying them access to high-profile positions in the industry's infrastructure or by restricting them to traditional feminine work spaces. Park authorities attributed these practices to one or more of the following reasons: women are not strong enough to handle the physical training required for certain positions, women are not ambitious enough to compete with men for these positions, motherhood is an obstacle, and other mountain professionals do not respect women's authority on the mountain and will not work well with them.

Mendoza and Aconcagua Adventure

The problematic representation of women in the base camp museum and their limited employment on the mountain echoes the way that Argentines as a whole were positioned in subordinate roles in early mountaineering enterprises. For much of the early period of Aconcagua mountaineering local participation was a working rural class endeavor and confined to the labor of the *arrieros*, or muleteers, who assisted foreign guides and transported supplies and equipment for the expeditions. The class disparity between the Argentines and Europeans on Aconcagua at this time left the *arrieros* out of the official history of Aconcagua adventure. It is this traditional interpretation of travel as a marker of cultural status (Clifford 1997: 33) and the concomitant understanding of adventure as requiring the economic and individual independence "to know with confidence what is intrinsically rewarding" (Sutttles 1983: ix), that posits the history of Aconcagua adventure within a classist framework that effectively "feminizes" the experiences of the *arrieros*.

On one level the belief that the *arrieros* have focused more on remuneration than essential life experiences has placed them on the lighter side of the balancing act between intrinsic value and extrinsic reward in mountaineering. This competition of incentives has often helped frame mountaineering's degree of adventure-ness. Suttles warned of the growth, in past decades, of those externally driven and corrupting impulses of commericialization and commodification in other sports. For Suttles mountaineering should ideally be measured in terms of personal growth as a "figurative experience, a metaphor in actual motion and hardship ... [that] revivifies for us a primordial experience of intrinsic gratification and sets a standard against which other experiences can be judged" (1983: x–xi). Suttles's cathartic expectations of mountaineering are analogous to Simmel's take on adventure: that it is extraneous to our everyday existence, yet it is existentially tied to our experience of life as a whole (1971: 188). In a traditional understanding of mountaineering adventure both Suttles and Simmel find resonance. Western cultural constructions and ideological underpinnings of mountaineering are found in Suttles's opposing of existential affirmation to capitalist commodity and in Simmel's reading of the outside as a formal aspect of the internal. However, when the internalized of mountain adventure faces scrutiny we find, as Ortner and Bayers have, that the experience of mountaineering is imbued with Western traditions of imperialism and individualism that have excluded non-Western subjectivities and have inherently assumed that adventure's natural state is marked as masculine.

The feminization of adventure, in this paradigm, would, by necessity, be that which the masculine defines itself against and could be enacted by the reversal of dominance of the paired components that Simmel uses to characterize adventure. An essential duality of adventure, according to Simmel, is the pairing of activity and passivity. Ann-Mari Sellerburg explains that for Simmel adventure "must embody an active mastery of *as well as* passive surrender to circumstance" (1994: 77). However, in the synthesis of these two ways of experiencing Simmel assumes that the dominance of active mastery is naturally an attribute of the masculine. Simmel uses the erotic as prime example of how adventure functions and argues that the degree of balance between those forces that give adventure its definition, this synthesis of activity and passivity or "conquering force and inextortable concession can, perhaps, be found only in the man" (1971: 195). Thus, for Simmel, women are excluded from the adventure of love, and by extension, from adventure in general, because "the activity of woman is typically permeated by the passivity which either nature or history has imparted to her character." (Simmel 1971: 195)

Therefore, the *arrieros'* "passivity," vis-à-vis their assumed lack of authority and initiative and "supposed" surrender to the expeditionary leader's
orders and circumstances, was to experience adventure from a "feminized"
position. Going one step further, this "feminization" of the *arriero's* adventure conforms to the discourse of imperialism theorized by Edward W. Said
in which Western masculinities are maintained through the subordination
or "feminization" of the Other (Said 1978). In this way the *arrieros'* feminized position stabilized the imperialist order inherent in mountaineering's
origins and whose legacy has contributed to the lack of recognition of the
Argentine *arrieros* in early endeavors on Aconcagua. Nevertheless, Western
adventure as it has evolved on Aconcagua would not have been possible
without the *arrieros*. Their exclusion from its history is a consequence of
mountaineering's impulse to maintain its traditional North-South hierarchies as well as its masculinist notions of adventure, or in other words, to
prevent slippage in the balancing of Simmel's dualities and the undoing of
Suttles's oppositions that would allow such hierarchies to topple.

For Argentines getting a foothold on mountaineering adventure has
generally meant accepting its traditional order and climbing up its classist
hierarchies via the routes of "imperialist masculine ethos." Not surprisingly, it was the Argentine military that followed in this tradition and mo-
nopolized local mountaineering until the late twentieth century. The growth
of Mendocino middle-class participation in mountaineering only came into
play with the redemocratization of the nation at the end of Argentina's military dictatorship, *El Proceso de Reorganización Nacional* [Process of National
Reorganization] (1976–1983). The creation of the Aconcagua Provincial
Park in 1983 democratized adventure on Aconcagua through civilian governance and an open admissions policy. With easier access to information,
equipment, technology, and the mountain itself, a new generation of world-
class *andinistas* from Mendoza appeared in the mid 1980s. It was the initiative and experience of this generation of mountain men who, in the style
of Euro/Anglo mountaineers, opened up new routes, set records on Aconcagua, and proved that mountaineering on America's highest peak was accessible to Mendocinos. These men are, in every respect, the founders of
the Aconcagua climbing industry, not only in terms of entrepreneurship in
offering mountaineering support to international climbers, but also in creating an educational and professional infrastructure for the formation of
high-altitude mountain guides in Mendoza.

Today the vast majority of Mendocinos on the mountain are not sportsmen, but employees or entrepreneurs in the mountaineering industry. In terms of class status, motivation for climbing, and travel they do not fit the traditional elitist profile of international mountain adventurers. Their presence is not just about seeking adventure as an activity of leisure and intrinsic reward, but also about engaging the adventure of economic risk that has opened the mountain to a much wider cross-section of local society. Nowadays, in addition to the Aconcagua *arrieros* employed to transport supplies to the two base camps, Mendocinos work there as porters (often students or guides in training), basecamp workers (many are teachers and students on summer break, some are ex-park rangers), guides, doctors, and the rescue squad (a special corps of the Mendoza Police Force). For them adventure in the mountain has been closely linked to enduring the physical challenges and climatic uncertainties necessary for earning a living. This does not mean that they do not view the mountain as a site of spiritual growth or personal transcendence. However, while on Aconcagua, the testing of personal limits, of risking all, burning bridges, and stepping into the mist, as Simmel has it (1971: 194), are not their only motivations for mountaineering and not necessarily the primary ones. Their synthesizing of what Simmel proposes as the foundational bipolarisms of adventure: chance/necessity; activity/passivity; certainty/uncertainty; is necessarily accommodated to the parameters of job requirements, expeditionary practices, and park regulations.

The differing approximations of locals and foreigners to mountain adventure are played out, nevertheless, within the selfsame legacy of the imperialistic and classist history of mountaineering. In this adventurous race Mendocinos still are at an assumed disadvantage due to the stigma associated with mountaineering for pay and the inference that they experience less danger, fewer risks, reduced independence, or in other words, "feminized" adventure.

Guides and Status

It is especially the Mendocino mountain guide who counters the notion of the local experience of adventure as being subordinate. By invoking Euro/Anglo modernity based on practices of safety, leadership, and professionalism, the guide claims a "higher" degree of adventure for Mendocinos than that previously bestowed upon local mountaineering support. He (and it generally is a he) shares with the international team members their expectations of adventure. While neither foreign, nor coming from a great distance, the guide is, nevertheless, the supreme adventurer of the group. He finalizes all major decisions about expedition movements; he deals with the uncertainty of group dynamics and weather, manages chance occurrences as they happen, and minimizes risk for all members of the group. (Of course, expeditioners still must contribute individual physical effort and commitment, but within the more controlled and limited environment provided by the guides' tutelage.) In speaking with Mendocino guides they have told me that they believe that the responsibility of their position marks them with privilege and status that is sometimes greatly exaggerated by expeditionary members. In fact, some guides have explained to me that at times expeditionary members see the guide in the guise of *"un ser increíble-un superman"* who transcends physical and spiritual limits. It is this idealized image of the local guide, his actions and reactions, his fortitude and knowledge, his personal sacrifice for others, that contradicts assumptions of exclusive international privilege of mountaineering, that ruptures the hierarchy formerly based on traditional First World–Third World dynamics of mountaineering expeditions, and that mitigates the stigma of his work for pay. Accordingly, the growing hegemony of the Mendocino guides challenges the cultural tradition of Euro/Anglo "heroic imperial masculinity," as Bayers calls it. Although the idea of "heroic masculine" adventure on Aconcagua has become a commodity sold by adventure tourism agencies to mountain enthusiasts, its real practice on the mountain is "feminized" for international clients by their subordinate relationship to local guides.

This higher status of Mendocino guides is further increased by international recognition of their achievement in mountaineering. This has especially bolstered local guides' status at home and increased mountaineering's visibility in Mendoza itself. Mendocino guide Heber Orona's completion of the Seven Summits (2006) and being named Mendoza's sportsman of the year in December 2000 signaled the preeminence of a new generation of Mendocino-trained guides and cemented the figure of the mountain guide into the Mendocino imaginary as a high-profile archetype with local and international connections.[4] However, linking mountain guides to Mendoza's affirmation of itself and its view of present and future potential begs a series of tough questions about gender-based notions of adventure and regional representation. Envisioning the symbolic value of mountain guides in this manner becomes greatly problematized when women enter into these roles that are taken to epitomize models of masculinity as well as national and regional identity.

Women and Adventure

La mujer es lo mejor que le pasó al Aconcagua.... Estás perdido en la montaña, saturado de hombres y de riesgos y una chica te sirve la comida, es maravilloso ... a los gringos les encanta. [Women are the best thing that has happened to Aconcagua... You're lost in the mountain, saturated with men and risks and a girl serves you a meal, it's marvelous, the gringos love it.]

> Rudy Parra, *Los Andes*

The mountain truly is a mirror; one's perceptions of the mountain reflect one's own self-perceptions.

> Thomas E. Taplin, *Aconcagua: The Sentinel of Stone*

Mendocino women taking on the role of mountain guides, however, is still not a common practice. Although nearly 30 percent of the student body studying at the Mendocino Escuela de Guías is female, there are few Mendocino women working on Aconcagua, even as assistant guides. While women are concerned about this disparity and have sought ways to create gender-based solidarity, female students of the school have not been supported by faculty in their attempt to form women's groups. They have been told by male instructors that there is no gender discrimination at the school and therefore no need for women's associations. While such may be the case in the classroom, the subsequent employment records and practices of the adventure tourism industry in Mendoza would dispute this notion of the lack of gender bias in the mountaineering industry, as would the traditional historical assumptions about mountaineering itself.

Although there are many international women climbers who come to Aconcagua (almost 12 percent of summit permits are issued to women and approximately 16 percent of all entrants to the park are female),[5] within Mendoza-based institutions and service providers women are either absent or not in positions of prominence. During the 2002–2003 climbing season there were no women in institutions of park authority; there were no women park rangers, doctors, or members of the rescue squad. In addition, there were very few Argentine women, like Mendocina Nancy Silvestrini,[6] who served as principal guide on an expedition. Even when women are accredited to guide they often find agencies unwilling to put them in positions of authority over male assistants. I know of one case where a female guide figured on official documentation as principal guide but she was not given the charge of the group nor paid as principal. One of the other male guides was given that responsibility and the accompanying higher salary, although he did not have the official credentials to be principal guide that his female colleague possessed. Nevertheless, the agency chose to put him in charge and she was left legally responsible, but with little authority to make decisions in her real status on the expedition as assistant.

Most of the women who pursue studies in the School and who do make it to the mountain, work, not generally as guides, but rather as supervisors and assistants in the cook tents at base camps. These positions, like all activity on the mountain, were originally male, but have more recently developed into a viable, yet not exclusive, space for women. This is not to say that theirs is a totally dependent and subordinate position on the mountain. Base camp workers do carry some degree of authority and power as they regulate the movement of goods up and down the mountain, provide a base of communication and safety for expeditions, store and inventory gear and supplies, ensure their employer's compliance with park regulations, and collaborate with park authorities, all in addition to preparing meals.

Symbolically domestic places limit women's mobility and restrict their access to risk and danger. Although women tend to be stationary, anchored for months at a time in one of the camps, they do constantly negotiate novel and unexpected situations on a daily basis that require spontaneity, ingenuity, and jumping headfirst into the breach. I have seen them find transport at a moment's notice for an injured or sick expeditionary member, set up a temporary camp for an expedition that inadvertently left half of its gear at the staging area below, find a way to send up replacement stoves or food to an expedition at high camp on receiving an unprogrammed request, lend food and supplies to alleviate a neighboring agency's supply crisis, notice symptoms of pulmonary edema in a client and send for the doctor, and celebrate the revelation of a client's, guide's, or porter's birthday at the last minute with a homemade birthday cake. This is not adventure in the traditional masculinized sense of mountaineering that presupposes movement, strength, physical sacrifice, peril, and the idea of spiritual quest. However, it is adventure that entails intense activity to navigate the unknown, to plan for the unforeseen, and to remedy the impossible. This is adventure grafted onto the necessary and daily routines that in themselves are complicated and often chaotic. With respect to adventure, then, there is both a parallel and a conflict between how adventure is actually experienced for Mendocino women who work on Aconcagua and their expected role in a masculine model of mountaineering adventure. Women on Aconcagua do much more than just fulfill the supposition that women's naturalized place on the mountain is the symbolic home where she merely serves up hot meals for the returning hero.

As for understanding mountain guides to be symbols of masculine and regional identity, Mendocinas' presence in this capacity, while still very limited and restricted, nevertheless, problematizes the equation. Their presence, however limited, disrupts masculine hegemony in the guiding profession and in so doing may also reflect a changing Mendoza society and social imaginary. Since the 2003–2004 season there have been more women on the mountain than ever before and with it the largest number of Mendocino women guiding in the history of Aconcagua. Opinions in the local Aconcagua industry hold that this has been due to the greater number of climbers and the lack of qualified and experienced guides in the 2002–2003 season to handle its 20 percent increase and the additional 15 percent increase in 2003–2004 (Garcia 2004).[7] By 2003–2004 even the park rangers employed at least two women to work at the entrance to the park (but not in the high camps) in 2003–2004. It appears, then, that in a repressed economy the boom of adventure tourism has necessitated the easing of access to heretofore gender-restricted positions that may benefit many young women today and in the future with aspirations to work or guide in the mountains.

Also indicative of a changing Mendocino consciousness was the media's attention to women's activities on the mountain during the 2002–2003 and 2003–2004 climbing seasons. The local papers emphasized international women's achievements by spotlighting, in feature-length articles, the youngest woman and the oldest woman on Aconcagua (Bruno 2003). In this same vein, high-profile stories in local press questioning the Mendocino rescue squad about the absence of women among its ranks brought to the forefront the underlying conservatism of Mendocino gender-based traditions. The challenge to the rescue squad's hiring practices in a feature article of *Los Andes* (Walker 2003a) may allow for an indirect critique of Mendoza social norms in general by removing a discussion of women's rights to a criticism of gender-biased practices on the mountain. It provides, in this manner, a safer way to address an emotionally charged issue by distancing it from the province's capital city. While this may insinuate the real and symbolic distance of Aconcagua from Mendoza, paradoxically it also recognizes the mountain's integral connection to how Mendoza envisions itself.

The Adventure of Aconcagua in Mendoza

Despite the snail pace of ascent, adventure on Aconcagua is still a heady experience when one considers that it is complex, multifaceted, transnational, and conditioned by both the past history of international mountaineering and the present-day exigencies of the business of tourism. The legacy of imperialism and its hierarchies of participation based on nation, class, and gender ruggedly endure on the mountain, just as a new sense of Mendocino adventure struggles to develop there within this selfsame tradition and in opposition to it. In this respect there is an incipient revaluing of Mendoza's history on the mountain that begs an understanding of adventure as not restricted to an elite, international leisured class and that would celebrate local *arrieros*, as well as the later, well-known Argentine military heroes, as mountaineering pioneers. There is also another reading of adventure that posits Mendoza's presence on Aconcagua within a globalized tourism market that discounts extrinsic reward as a corrupting factor in mountaineering and projects the Mendocino professional mountain guide as superlative adventurer.

In returning to the T-shirt slogan that started this essay, *"toda adrenalina"* seems to describe the dynamic nature of adventure that is evolving and mutating on Aconcagua as it adapts to the needs of the industry and satisfies the desires of international climbers. The adrenaline rush that drives the economics of the Aconcagua industry pushes Mendocinos to jockey for position in an international market and has instigated a process of democratization and new feminizations of adventure.

If, as Ortner suggests, mountaineering has been about the defining of the masculine self, then it tends to follow that women's presence on Aconcagua would find the best footholds in those interstices where feminine activity would not destabilize this process. Such has been the case, and subsequently, the latest feminization of adventure, not in the sense of a "less masculine," lesser, or subordinate experience, but in a reevaluation of the feats, accomplishments, and undertakings of women on Aconcagua, necessitate a more complex understanding of how adventure is perceived, experienced and constructed. The feminization of adventure, in terms of women's activities on Aconcagua, may still tend to follow patriarchal patterns of gender-based territories and labor, but it also, paradoxically, disputes a single definition of adventure grounded in the competition for masculine supremacy. In this respect, the visibility of Mendoza's female mountaineers working in all capacities on Aconcagua, supported by the local media, has begun to draw attention to the inequalities of gender divisions on the mountain and has begun to call into question the supposed naturalness of the assumptions on which they are based.

It is within this complex and multilayered context that I discern adventure on Aconcagua as daring and enterprising in the way that it characterizes a Mendoza striding into the twenty-first century, "knowing with confidence what is potentially rewarding for the future" (to paraphrase Suttles), as it explores its options within the adventure tourism market of a globalized world. From this viewpoint Mendoza's ventures on the mountain embody adventure on Simmel's terms: while mountaineering takes place outside of the capital of Mendoza and is seemingly extraneous to the normative customs of Mendocino agricultural heritage, its intense, risky, and uncertain evolution is intrinsically linked to Mendoza's economic, social, and cultural makeup and well-being. The sometimes unnerving and thorny issues of the mountain are now, as Grajales would have it, integral to Mendoza itself.

Notes

1. All translations are my own.

2. The first European attempt at summiting was in 1883 by Austrian geologist Paul Güssfeldt, and the first successful recorded summit was accomplished fourteen years later in 1897 by Matthias Zurbriggen, the Swiss guide of an English expedition under the leadership of Edward A. Fitz Gerald, who chronicled it in 1899 in *The Highest Andes.*

3. I have written elsewhere about the one new feminine garment added in 2005.

4. Orona was named the outstanding Mendoza athelete (over stars of soccer, rugby, volleyball, swimming, cycling, boxing, basketball, and car racing) by the Circle of Sportswriters and was also given top honors for the year by MÁS Sports, a division of *Los Andes.*

5. According to 2001–2002 data available at the Aconcagua Provincial Park's official website: http://www.aconcagua.mendoza.com.ar

6. Silvestrini was killed in July 2003 in an accident in the Himalayas. She had been the first Argentine woman to summit an eight thousand meter mountain. Caught in a storm on her descent, she fell to her death and her body has not been recovered.

7. Admission continues to grow. 2005–2006 saw a 12.5% increase over the previous season.

Chapter 11

The Toughest Job You'll Ever Love: The Peace Corps as Adventure

Michael J. Sheridan and Jason J. Price

"It is in the compelling zest of high adventure and of victory, and in creative action, that man finds his supreme joys."
—Antoine de Saint-Exupéry

For the modern American in search of a state-sanctioned adventure there is always the Peace Corps: that far-off other world where the privileged and daring set off to do good and find themselves. More than 170,000 Americans have served in the Peace Corps since the program's inception some four decades ago. The authors of this essay rank in those numbers. Sheridan served as a water technician in Kenya from 1988–1990. Price was a secondary school teacher in Malawi from 1999–2001. We both joined, in part, out of a search for adventure. The Peace Corps provided us with the means to balance our personal desires with moral purpose, to experience adventure in the name of both ourselves *and* the world. This impulse is hardly uncommon, nor is the association between Peace Corps and adventure. This essay, then, explores the notion of Peace Corps service as "adventure" by drawing particularly upon the formal sociology of Georg Simmel.

"Adventure" has long marked the shape of Peace Corps service. An early slogan, "the toughest job you'll ever love," linked extremes of effort with extremes of emotional fulfillment. Later, the agency posed the question, "How far are you willing to go to make a difference?"—linking extremes of commitment with extremes of impact. Now Peace Corps simply issues would-be volunteers the challenge: "Life is calling. How far will you go?" Despite these subtle shifts in emphasis, the overall point is clear and consistent: Peace Corps is a difficult and challenging experience that offers innumerable social and personal rewards—an adventure of the highest moral purpose.

A problem arises, however. The Peace Corps Volunteer (hereafter, PCV) does not set off on an entirely perilous and wholly unpredictable journey. She does not throw herself mercilessly into the current of the world. She is not some mythical adventurer who blazes a trail into the unknown like Ahab, Tarzan, or Ulysses. Instead, the PCV walks a familiar path with clear signs and a reliable support staff. She is salaried and insured. Her conclusion is imaginable. This contradiction begs some questions. With all of the term's connotations of risk, mortal danger, and the exotic, how can such a well-established, overdetermined government program like the Peace Corps be so consistently marketed and perceived as an adventure? Does the Peace Corps experience even constitute an adventure? What might be the advantage of understanding Peace Corps service as an adventure?

Simmel, the brilliant and original, if often opaque, German social philosopher, spent the better part of his career identifying, carving out, and polishing different forms of social interaction from the rich and variegated canvas of human social life. Examples of these forms include "exchange," "conflict," "domination," "prostitution," and "sociability" (Levine 1971). These forms are abstract and unreal, significant only as analytic tools. Simmel maintained that a working knowledge of social forms is invaluable because it confers upon the analyst an ability not only to see social interactions more deeply, but also to draw broader, more intriguing connections across diverse social planes. This, he argued, is the essential goal of the sociologist.

"The adventure" is a distinctive form of experiencing the world. It has a specific beginning and end. It occupies a place both *in* and *out* of the adventurer's life. To Simmel, the adventure represented "an island in life … bound to the consistent life-process by fewer threads than are ordinary experiences" (Simmel 1959: 244). It is the form of the adventure that impresses its own particular pattern on social interaction. This form includes the antagonism, compromise, and synthesis of three core oppositions: activity and passivity, chance and necessity, and certainty and uncertainty. This leaves the adventurer in a rather precarious position. She must conquer her experience *and* surrender herself to it. She must transform unexpected swings of fortune into integral elements of her journey. And she must endlessly redefine her goal in order to maintain a forward momentum of constant risk.

We believe that by looking at Peace Corps service through the lens of "adventure" we will be able to understand the experience more deeply. To that end, we begin with a discussion of the conflicting motivations for joining the Peace Corps. We then analyze the antagonism between Simmel's three core oppositions and what techniques PCVs utilize in order to resolve them. These oppositions arise regularly throughout Peace Corps adventures because two contradictory patterns of symbolic work are mapped onto the social organization of every service. On one hand, there is the institution's model of the adventure which is often defined by activity, necessity, and certainty. On the other hand, the volunteer's experience of adventure is regularly characterized by passivity, chance, and uncertainty. As a result, the PCV must negotiate the difficult tensions between the ideal and the real, form and content, model and experience. The symbolic work of creatively synthesizing frustrating oppositions into neat, symbolic resolutions will mark the basis of our analysis of Peace Corps service as adventure.

The Self and the Selfless

The Peace Corps adventure begins long before the PCV sets off for some proverbial, out-of-the-way village. In order to qualify for service, every applicant must interview and submit a short statement of purpose. Central to both tasks is the applicant's motivation statement, which is often reducible to one simple and straightforward declaration: "I want to join the Peace Corps to help people." Price recalls his 1999 interview:

> The recruiter had served in the Malian Sahara years ago and the experience had changed his life, he said. He asked me why I wanted to join. I told him that I felt the need, the obligation, to give something back to the world. Whatever talk there might have been of exploration and discovery in joining the Peace Corps, I always returned to this point—my sense of obligation to serve humanity. And he would nod approvingly and jot down notes on my application.

Most applicants do want to help people, but this can be an oversimplification. Family and friends invariably ask why a PCV should go so far when there is no scarcity of people needing help closer to home. The answer, of course, is that most are lured as much by adventure as by goodwill. Many applicants do want to help people, but most are also looking to set off for an "island in life" with boundaries that distinguish the remarkable from the ordinary. They desire something greater for themselves beyond the scope of their everyday experience.

Hamilton has argued that self-interest is essential in any discussion of adventure. "The adventure itself can be undertaken for one's own personal perhaps secret purposes rather than for society's benefit; and it holds the promise of a rapid, if not an easy, road to achieve one's particularistic goals" (Hamilton 1978: 1478). Similarly, many applicants to the Peace Corps also do so to "make a difference" in the world, but also for their own secret purposes and particularistic goals. This highlights a tension that confronts every Peace Corps applicant—the tension between the selfish and the selfless, between altruism and egoism.

The Peace Corps does not deny the gravity of this opposition. It is addressed immediately in the recruitment literature in the section, "Why should I volunteer?"

> You give and you get. The chance to make a real difference in other people's lives is the reason most Volunteers serve in the Peace Corps.
>
> But that is not the only benefit of Peace Corps service. Volunteers also have the chance to learn a new language, live in another culture, and develop career and leadership skills. The Peace Corps experience can enhance long-term career prospects whether you want to work for a corporation, a nonprofit organization, or a government agency. The Peace Corps can even open doors to graduate school.
>
> From practical benefits such as student loan deferment, career benefits like fluency in a foreign language, and the intangible benefits that come with making a difference in people's lives, there are a variety of rewards for the dedicated service of Volunteers. Rewards that last a lifetime. (Peace Corps 2004)

This emphasis on moral reciprocity—"you give and you get"—marks an attempt to synthesize these frustrating oppositions and allay any existential anxiety before it takes root in the applicant's mind. Certainly the PCV "gets" in the form of language ability, career and leadership skills, career prospects for private corporations or federal agencies, and increased chances of acceptance into graduate school. But it is essential, also, that the PCV "gives" and makes "a real difference." By deftly mingling the selflessly altruistic with the selfishly pragmatic into a coherent synthesis, the recruitment literature attempts to resolve this tension.

Easy resolutions offered in recruitment brochures fail to solve this problem, however. In fact, many PCVs find themselves wrestling with this tension all throughout their service. The symbolic work of the application process, then, foreshadows other symbolic oppositions that surface throughout the service itself: namely the tension between the active and the passive, chance and necessity, and certainty and uncertainty. These are the tensions that Simmel used to define the form of experiencing known as "the adventure."

The Active and the Passive

Peace Corps constructs an image of the PCV as an activist adventurer that is opposed to the passive role he often assumes in his community. It makes for a frustrating opposition. As Simmel puts it, "on the one hand, we forcibly pull the world into ourselves" and yet on the other "we abandon ourselves to the world" (Simmel 1959: 248). In Peace Corps service, as in all adventures, there is a complex ebb and flow between the active and the passive, "between what we conquer and what is given to us" (1959: 248).

The PCV as activist adventurer is an historical construction rooted in President Kennedy's Cold War vision of "New Frontiers." Sargent Shriver, the first Peace Corps director, organized the fledgling agency around the notion that PCVs would spread the gospel of development to the backwaters of the world and counter the imperialistic image of the "ugly American" in the process (Burdick and Lederer 1958). His vision was broad, ambitious, and comprehensive. He said he hoped to find "the rugged Americans needed for this pioneer job [who would] work clearing the jungle of eastern Andean slopes, building farm houses, improving crops, starting education, and surveying for access roads from the jungles to the larger centers" (cited in Fischer 1998: 19). Here was a modernizing adventurer: a stalwart individual who was innovative, stubborn, and highly motivated by notions of democracy and populist justice; a modern analogue to the mythic nineteenth-century frontiersmen and cowboys that dominated the American television shows and movies of the 1950s and '60s. Shriver even went so far as to describe this ideal in sexual and militaristic images—"The point of the lance is lean, hard, focused. It reaches its target" (Shriver 1964: 5)—calling to mind Hemingway's manly and domineering depictions of adventuring (Hoffman 1998: 198).[1]

The U.S. Postal Service's 1999 "Peace Corps Stamp" demonstrates this ideal. The image shows the prototypical PCV at work in the field. The industrious PCV, illuminated by a shaft of white light, explains the mechanics of a plow to some attentive Africans. He stands at the center of the image and actively instructs them. They stand at the periphery and passively learn

Figure 11.1 • US Peace Corps postage stamp.

from him. His pose signifies both physical and social command of the situation, while their pose signifies a certain primordial approbation. Though clearly in charge, the PCV remains a servant of the people. He is Shriver's "point of the lance" moving toward the target of development, carrying with him the enlightenment of American know-how. He is, to quote PCV Paul Theroux, "the most fortunate liberal on earth" (Theroux 1998: 57).

In their activism, PCVs are encouraged to embody this image. But the lived world of Peace Corps service is very different. In fact, PCVs often find themselves the passive recipients of ecological, institutional, and cultural conditions that limit their actions. Community development volunteers work alone amidst complex and dynamic host societies, largely ignorant of the intricate mechanics of social change and the power of local tradition. Teachers often find themselves teaching by rote for national exams rather than addressing the deeper needs of their students. Environment volunteers often spend months determining if the job they were sent to do can even be done. The institutional vision of the PCV as activist adventurer is a "development narrative" that defines a problem, a justification for intervention, and a solution (Roe 1991). But the lived experience of the PCV rarely includes this narrative format, moral purpose, and optimism. As a result, volunteers quickly realize that they are not some hybrid of Davy Crockett and John F. Kennedy. Instead, many bear a greater resemblance to Simmel's well-known social type "the stranger"—that person "who intrudes as a supernumerary, so to speak, into a group in which the economic positions are already occupied" (Simmel 1950b: 403). Price recalls his socialization into this frustrating reality:

> There is a practice common to some African schools that I first found scandalous. I remember the first time it happened. My headmaster arrived late. He had been busy finalizing "our appointment." I had no idea what this meant. I soon learned that it referred to a terminal round-up of the school's female population for a surprise pregnancy test at the maternity clinic only 300 meters up the road. We would announce the appointment at the morning assembly, separate the girls from the boys and march them up the dusty road for their examinations. The girls found pregnant would be immediately dismissed from school, along with anyone who dared flee the scene.
>
> I was appalled and protested immediately, "How can this be true? We must respect our students' privacy! This is an infringement on these women's rights!" But my protestations were only met with good humor, even from the most liberal of faculty members. Mr. Banda, a well-mannered English teacher and Presbyterian minister, broke in diplomatically, "It will be alright, Price, you will learn. As for today, feel free to take the boys down to the football pitch if you wish. In fact, feel very free. Do whatever you wish."
>
> I stayed and taught the boys. And as they took notes off the blackboard, I watched as the girls walked up the dusty path to the clinic in their white blouses and royal blue dresses. Two of them were discovered to be pregnant. Another tore her dress hopping a rusty fence in escape. She had been seven months pregnant and hadn't showed in the slightest.

Intense frustration is often born out of the misfit between the institutional adventure model and the volunteer's reality. Many PCVs fail to embody Shriver's iconic adventurer who patiently demonstrates American ingenuity to appreciative third-world villagers. Instead PCVs often become the passive objects of institutional role-making or other people's strategies. Sheridan's PCV career illustrates these tensions:

> Peace Corps sent me to Kenya to be a "water tech," so I joined a group of seven water engineers to learn about building cement tanks for collecting rainwater from metal roofs. After three months learning the basics of masonry, Peace Corps assigned me to a town where few people were interested in water tanks. There were many natural springs in the area for capping and pipeline development. Every village had a water project committee, and my uncomfortable role was to survey and design the pipelines and then solicit money from an American development agency active in the area. After teaching myself rudimentary survey techniques, frictional headloss parameters, and drafting methods, I completed nine proposals, solicited funding, and built two gravity-flow pipelines with evenly spaced public water taps. I reckoned myself a moderately successful PCV.
>
> Within a year after I left Kenya in 1990, villager leaders wrote me to describe how these projects had fallen apart. Soon after I had left, the wealthier uphill villagers had tapped into the main pipeline to bring water into their houses. The increased friction meant that the water could not reach the midline storage tanks, so the downhill pipes became dry. The project committees bickered over who was to blame while the intakes and pipes became completely plugged up with silt. I realized that I had been both an ignorant stranger and a community insider.

Volunteers can be ignored and manipulated just as much as they can aid and assist. In an unexpected twist of fate, they become more like Simmel's strangers than Shriver's saviors.

To resolve the dissonance between the Peace Corps ideal and the volunteer's reality without slipping into cynicism, many PCVs craft adventure narratives. By telling stories about their resiliency, adaptability, and good humor when facing situations beyond their control, PCVs synthesize active and passive roles in order to re-imagine themselves in light of these models. These "Peace Corps stories" provide narrative frameworks for troubling Peace Corps experiences. They are told, retold, cultivated, reified, and eventually co-opted by other PCVs. By the end of her service, each volunteer has a veritable catalog of narratives about frightful diseases, mind-blowing

injustices, and totally absurd encounters. These stories are valuable because they convey otherwise complex, existential dilemmas with eloquence
and economy. Taken as a whole, this corpus of "Peace Corps stories" comes
to represent the central mythology of a PCV's experience, occupied as it is
by the familiar characters, predicaments, joys, triumphs, and absurdities of
everyday life. Lévi-Strauss would find these myths "good to think" because
they help the PCV to structure the chaos and ambiguity of volunteer service (Lévi-Strauss 1966).

A story that articulates the creativity involved in synthesizing the opposition between the active and the passive concerns Nathan, a PCV school-
teacher in Malawi who had accompanied his students in an open-air lorry
for an away athletic match. For many students, and some teachers, this
marks a rare opportunity for long-distance travel. Below, the PCV recounts
the first (and only) time he chaperoned one of these trips:

> We arrived at the host school in the early afternoon—85 strong—
> after an unplanned pit stop at a Christian mission along the way to, seri-
> ously, weld our flat-bed truck back together (apparently the lorry was
> cracking down the middle like a pistachio). The bumpy dirt road had
> taken its toll on the vehicle.
>
> The driver had picked us up that morning smelling of alcohol. Upon
> arrival, he almost hit three women on the sideline before swerving to a
> stop in the middle of the field in front of the host team. He hopped out of
> the cab and stumbled off to the nearest watering hole.
>
> When he came to pick us up after the match he could barely see
> straight. This worried everyone. But no one denied that he would drive
> us home. When I stated the obvious objections, my fellow teacher cut me
> off, saying "this is the way we do things here."
>
> Trying hard to be sensitive to their way of life I agreed to let him
> drive. Another more concerned adult asked me to sit next to him, figur-
> ing I could grab the steering wheel if we veered too close to a tree or goat
> or worse.
>
> But then everyone demanded to travel home via the nearby paved
> road—a much longer journey but without bumps.
>
> No way! No. No no no no no. This guy driving us back at night on
> the notorious main highway of Malawi! The road was barely wider than
> two cars and full of twists as it climbed its way back into the mountains.
> Nope nope nope.
>
> Cultural sensitivity was completely forgotten as I dragged the worth-
> less drunk out of the cab and refused to budge from the steering wheel.

I suggested that we spend the night.

"Where are we going to sleep? There are 85 of us."

I suggested that we take the dirt road.

"Absolutely not."

I suggested that someone else drive.

"None of us know how to."

The sun had just set. And something would have to be done.

Nervous, I drove the whole way in second gear. I started out on the right side of the road, which in Anglophone-Africa is the wrong side. Distant headlights coming towards us eventually reminded me to get the hell out of the way. I had to drag the side of our truck into the bushes to make room for the oncoming traffic, only to get an earful from the passengers in the back. They didn't appreciate that those bushes were a little more forgiving than the 30-ton, kamikaze trucks roaring through the darkness on their way to Tanzania.

About halfway home, a concerned driver forced me to stop, telling me that the truck—with 85 people on the back—was cracking in half lengthwise.

Almost humorously, the teachers told him that we knew all about it, so bugger off. Besides, we were in the middle of nowhere and had to get home.

Three hours later and back in our village I sunk out of the driver's seat and stumbled into my house. Before we arrived I was able to let off some steam by using every curse word I have ever known when the drunk complained that I was grinding his gears. I should have just burnt the clutch at the beginning. Everything would have been simpler.

[Source: personal communication in possession of the authors]

Nathan's amusing, yet disturbing account articulates the antagonism between the active and the passive that challenges most PCVs.[2] By "dragging the worthless drunk out of the cab and refusing to budge from the steering wheel" he performed his role as an activist adventurer. But he could perform that role only within the narrow and rather absurd confines of the cultural conditions that limited his scope of action. By crafting this troubling event into a compelling narrative and sharing it with other volunteers, the PCV "creatively re-imagined his situation and thereby regained mastery over it" (Jackson 1998: 23).

The path in unifying these opposites is challenging, but it is in this synthesis that lie the rewards of the adventure. Simmel writes that "it is among adventure's most wonderful and enticing charms that together our activity and our passivity—the unity which even in a certain sense *is* life itself—accentuates its disparate elements most sharply, and precisely in *this* way makes itself more deeply felt, as if they were only the two aspects of one and the same, mysteriously seamless life" (Simmel 1959: 249).

Chance and Necessity

Peace Corps implies that service will fit easily and seamlessly into the life course of its volunteers, yet so much of the experience is accidental and left to chance. This opposition highlights a misfit between chance and necessity, "between the fragmentary materials given us from the outside" and "the consistent meaning of the life developed from within" (Simmel 1959: 247). Peace Corps situates service within the pasts and futures of volunteers, implying a compelling logic and an undeniable relevance to service. The first page of a recent recruiting brochure reads, "Put your liberal arts degree to work ... people with liberal arts degrees often have a well-rounded education and a perspective that brings creativity and versatility to their Volunteer assignments" (Peace Corps 2002).[3] This implies a logical connection between the PCV's past and her service. In this light, Peace Corps service represents a logical and necessary experience in the life course of the volunteer.[4]

But Peace Corps service is all too often accidental and illogical, far removed from the continuity of a volunteer's life. This may begin as early as country placement. An applicant with mathematics training and fluency in French and Wolof was sent to Malawi to teach English and not Senegal to teach arithmetic. A trained writer with competency in Spanish was shipped off on a community development assignment in Kiev and not for ESL instruction in Lima. Chance continues to determine experience once service has begun. In a story recounted by Sheridan, it was nothing more than a meddling goat that marked the end of one PCV's service in Kenya:

> Jon and I were sitting under the buzzing fluorescent lights of a Mombasa bar one evening, swatting mosquitoes and trading frustrations.
>
> No one was supposed to graze their animals on the school compound, Jon said, but one old man kept bringing his goats there every day. A particular goat had made a habit of coming into his classroom every day

and disrupting class quite effectively. Throwing stones, complaining to the headmaster and local leaders, and talking to the old man had proved futile.

I responded (with classic PCV ingenuity, I thought) that he should use some whitewash and the black paint the school used to make black-boards to paint the goat like a zebra—a painless sanction.

Two weeks later I learned that Jon had innovated upon my idea. He had painted the old man's name on the side of the goat. The old man was so incensed at this outrageous public insult that he summoned his son, a professional wrestler, from Mombasa to expunge the shame. The wrestler and the teacher fought in the school compound during classes—to the delight and consternation of the entire village—and Peace Corps decided to send the teacher home in order to avoid a diplomatic incident.

Whatever the case, whether it plays a fundamental or minor role in deter-mining the PCV's experience, chance exists, ready to subvert the adminis-trative notion that Peace Corps service is logical and necessary in the life course of the volunteer. In the end, Peace Corps service is just as absurd as it is logical, just as unnecessary as it is necessary.

This is difficult for the PCV who has given up her old life in search for a meaningful adventure. She is, therefore, challenged to resolve the antag-onism between necessity and chance, meaning and meaninglessness. What makes this resolution even more difficult is that synthesis between chance and necessity often comes through demonstrations of cultural competence, outside of the formal social roles of the PCV's particular assignment. As a result, trivial experiences often become the most meaningful, whether it be the mastery of some local card game, joyful rounds of drinking, fluency in indigenous swear words, or a shrewd bartering session. These experiences demonstrate the volunteer's ability to transform unlikely opportunities into integral and necessary aspects of Peace Corps service. She symbolically as-sociates accident and coincidence as necessary parts of the central meaning of her experience (Wanderer 1987: 27), for it is these moments of creative improvisation that reveal one's cross-cultural know-how. In the process, she expands her notion of what is logical, necessary, and meaningful and em-braces some of the unpredictable, accidental, and seemingly meaningless elements of social life. The result, according to Simmel, is nothing less than a harmonious union of the internal self and the external world—"where the course of the world and the individual fate have not yet been differen-tiated from one another" (Simmel 1959: 250).

Certainty and Uncertainty

Peace Corps constructs an image of service as logically ordered and its development work as categorically positive. In reality, however, the Peace Corps experience is often unpredictable, and the complexities of development endow many volunteers with existential doubt. For Simmel, this tension between certainty and uncertainty, between "knowing an outcome" and being unsure of "arriving at that point for which we have set out," is fundamental to the adventure (1959: 249).

Peace Corps orders service by organizing the PCV career around rituals of separation, initiation, and reincorporation. This framework of certainty and predictability begins with a ritual of separation called "staging," in which a training group gathers in an American city for several days of orientation. The group then departs for their host country. After about three months of mental and bodily discipline during community based training (CBT), the initiates are officially transformed into PCVs by "swearing in"—taking an oath to serve the host country and the U.S. government. After one year of service, the cohort becomes a corporate group again during the "in-service training" (IST). Here the PCVs regroup to collectively reflect upon their progress. Finally, there is a close-of-service (COS) conference. Here PCVs assess their experience and prepare for "re-entry" into American society. In total, these rituals confirm an ideology that runs through all institutional activities—that is, the affirmation of development and modernization in the face of poverty and inequality. The idea that Peace Corps is an adventure of the highest moral order rests on this notion. And so, with lawful order imposed on a service endowed with moral certitude, the volunteer is rarely supposed to feel uneasy.

Despite the symbolic order encoded upon service, however, Peace Corps volunteers often feel very uncertain about the order of their service and the underlying assumptions that constitute its motive force. "Early termination" (ET) rates vary from country to country, but roughly one in four PCVs worldwide never completes their service. These cases can subvert the sense of order and certainty that Peace Corps diligently inscribes on service. Meant to reinforce a sense of certainty and reconfirm an institutional ideology, it is ironic that these rituals can have the opposite effect.

Many volunteers who actually do complete their two years of service may endure interruptions, such as medical or psychological evacuations or disruptive changes in field sites or work schedules. Political uprisings or shifts in global politics can even cause programs to close. Peace Corps Volunteers quickly become aware that an orderly structure for their experiences cannot ensure that they will actually complete their service.

Among some volunteers a deeper, more existential uncertainty develops throughout the course of their service. This concerns their general idea of development. For others it deals with their personal roles as grassroots development workers. Such reservations have been explored extensively and do not demand our comprehensive attention in this essay (see, e.g., Escobar 1995, Maren 1997, Hancock 1989, Robertson 1984, Terry 2002, Schwimmer and Warren 1993). In many ways, the PCV works in the ruins of the volunteers that preceded him, such as the rubble of a new science laboratory, innumerable copies of unread Faulkner novels in a school library, silted-up spring-boxes and empty pipelines, and cartons of expired condoms in health center storerooms. It takes time for volunteers to understand the complexities of development; the extent of their own ignorance regarding culture, history, and the mechanics of social change; and their own potential to do something. In many cases, by the time these lessons are learned, the two years of PCV service are nearly over. The certainty of development ideologies fractures under the pressure of experience.

The tension between certainty and uncertainty in the Peace Corps adventure is often the most troubling. Some PCVs return to America because they cannot make sense of their service; others become cynical and hole themselves up in nearby expatriate bars, coffeehouses, or nightclubs. The volunteers who do manage to keep going often do so through a redefinition of their goals (Wanderer 1987: 26). According to Simmel, this was a logical response: "When the outcome of our activity is made doubtful by the intermingling of unrecognizable elements of fate, we usually limit our commitment of force, hold open lines of retreat, and take each step only as if testing the ground" (Simmel 1959: 249).

The typical PCV, then, continuously scales down her goals from developing a country to developing a community, until she becomes satisfied with making even the tiniest change for the better. Sheridan rationalized his PCV experience with the idea that it was a net gain for Kenya if he prevented just one child from dying of a water-borne disease. Price stopped investing all his energies in his classroom teaching and focused instead on transferring his top students to better secondary schools. Eventually, many volunteers turn inward in order to focus on making changes in themselves rather than their host societies. In the PCV libraries that are omnipresent features of all Peace Corps offices, pop philosophy and self-help manuals far outweigh host country histories or ethnographies of development interventions. As Sheridan recalls,

> We devoured these books on Zen Buddhism and the healing power of prayer, I think, as a way to make sense of our increasingly uncertain sense of self by creatively adopting new cosmological frameworks. In our barroom bull sessions, we papered over the cultural contradictions we were encountering by debating the merits of Taoism. We had stopped trying to develop Africa and begun trying to develop ourselves.[5]

Why does this happen so consistently in Peace Corps programs throughout the world? In part, because the inward quest seems more manageable than the vast and complex forces that shape the local histories in which PCVs are embedded. It is a logical evolution that leaves the PCV in a most peculiar situation. She returns to America with a contradictory sense of her own agency in the world (Dunn 2004). One the one hand, she feels an overwhelming sense of empowerment, as if she can do anything, live anywhere. On the other hand, she comes to terms with her own impotence in the face of the world and its social problems. This marks yet another tension that begs creative resolution long after the official adventure is over; in many ways, a whole other adventure altogether.

Conclusion

Not every Peace Corps service is an adventure. Some are mere experiences. According to Simmel, an experience becomes an adventure only by virtue of experiential tension that is resolved in a space outside the usual continuity of life. In Simmel's words:

> The *content* of the experience does not make an adventure. That one faced mortal danger or conquered a woman for a short span of happiness; that unknown factors with which one has waged a gamble have brought surprising gain or loss; that physically or psychologically disguised, one has ventured into spheres of life from which one returns home as if from a strange world—none of these are necessarily adventure. They become adventure only by virtue of a certain experiential tension whereby their substance is realized (Simmel 1959: 253).

Peace Corps experiences are inclined to become adventures because two different symbolic patterns of meaning are mapped onto every PCV's service. On the one hand, the Peace Corps frames service as an ideal model built around notions of activity, necessity, and certainty. On the other hand, the PCV experiences a service grounded in situations of passivity, chance, and uncertainty. There exists, therefore, a jarring disconnect between form and content, ideal and experience, model and reality. This fundamental opposition constitutes the experiential tension that defines many Peace Corps experiences as adventures.

In this chapter we have explored these frustrating tensions, along with the creative techniques volunteers employ to resolve them. To synthesize activity and passivity many volunteers tell stories highlighting their resiliency in difficult situations. In coming to terms with chance and necessity volunteers expand their notions of what is meaningful by embracing experiences that showcase a certain cultural competence. Finally, in dealing with certainty and uncertainty, volunteers reevaluate their goals in order to ensure certain measures of success. With each of these oppositions comes a certain tension and with each tension comes a demand for a creative resolution. Some volunteers fail to realize these oppositions. Others ignore their residual tensions, and still some fail to resolve the antagonism between these oppositions. We cannot say that these volunteers have experienced adventures, no matter how adventurous the content of their experiences may have been.[6]

The adventure is marked by the struggle between opposing forces and the challenging experience of mastering them in a balanced, symbolic resolution. Its crux lies in its challenge, its value in the creativity that resolution demands. For that reason, the adventure is not a comfortable form of experiencing the world. It is a form shaped by tension, antagonism, and frustration. Perhaps this is why Simmel noted that "so much of life is hostile to adventure" (1959: 255). At its base, it is a radical social form, but a form that must be expressed, a form grounded in that entirely human wish to cross the boundaries of the everyday into the field of the remarkable. If nothing else, then, the adventure promises one thing: vitality. "Certainly, it is only one segment of existence among others," writes Simmel, "but it belongs to those forms which, beyond the mere share they have in life and beyond all the accidental nature of their individual contents, have the mysterious power to make us feel for a moment the whole sum of life as their fulfillment and their vehicle, existing only for their realization" (Simmel 1959: 258).

Adventure looms large in Peace Corps, and with good reason. To send volunteers (and to send one's self) spinning away from the continuity of life, tethered to its core by only a few threads, is adventurous. In such a situation—in which ideals and realities consistently battle one another in an unremitting, symbolic tug-of-war—one is likely to find such a summary of the human condition. But does one have to go to the ends of the earth to be an adventurer? Is this the only location where "the whole sum of life" may be found? Of course not. Though veiled by mundanity, adventure "lies latent in every experience" (1959: 256). They are ubiquitous, not unique. In everyday life, oppositions abound. Tensions exist and the world demands creative resolutions at all times, in nearly every experience. The adventurer need only acknowledge them as such, make the invisible tensions visible, and realize that the boundaries between the remarkable and the everyday are more blurred than she might have imagined.

Notes

Acknowledgements: The authors would like to thank T.O. Beidelman and Mwenda Ntarangwi for their perceptive comments.

1. It is significant that the original Senate bill for the establishment of the Peace Corps posited an all-male corps (Fischer 1998: 91). The term "corps," of course, reflects the institution's role as the symbolic opposite of the U.S. military in Cold War geopolitics.

2. Levine, paraphrasing Zajonc (1952) writes, "strangers are expected to conform to host culture norms and find those expectations disturbing owing to conflicts with values brought from their home culture, they will tend to express aggression against those norms" (Levine 1985: 78).

3. James Scott's analysis of state-led development interventions supplies the terminology for understanding the administration's response to the structural vulnerability of institutionalized chance. In order to maximize the bureaucratic "legibility" (Scott 1998) of the volunteers and be able to impose its own vision of development problems and solutions on these PCVs, the Peace Corps orients its recruitment efforts toward college graduates without highly focused technical skills, area studies knowledge, language skills, and international experience. These so-called "B.A. generalists" are relatively blank slates—and therefore "legible"—targets for Peace Corps training to inscribe with particular forms of knowledge, values, and theories of community development.

4. The administration also places a high value on chance, which is a vital aspect of the ideal volunteer career of innovation on the New Frontier (Fischer 1998,

Hoffman 1998). Throughout Sheridan's three months of Peace Corps training, teachers preached the virtues of flexibility and chance. If, for example, a volunteer's water project was not turning out well, but she happened to receive a basketball as a gift from home, it was a fine time to build a basketball court.

5. It is unclear how this process of nascent, but diffuse, "vision questing" among current and past PCVs relates to the historical emergence of a "self-help" literature in American society in the 1970s and 1980s. The Peace Corps experience is an intensely emotional one, and many PCVs of the authors' acquaintance have struggled to express how the experienced framed some of their highest emotional highs and the lowest lows. The PCV struggle to reconcile Simmel's existential oppositions may mean that after two years of fieldwork, many PCVs are quite as culturally disoriented from the tensions and contradictions of adventuring as the doped-up explorers of Central Africa (Fabian 2000). These tensions are, of course, as fundamental to anthropological fieldwork as development volunteerism.

6. The compulsion to continue the symbolic work of the Peace Corps adventure may explain why so many returned PCVs synthesize their bicultural identities into lives of public service. Between 1961 and 1995, about 25 percent of 145,000 returned PCVs became teachers or academics, and another 10 percent entered government work with foreign postings. As of 1995, about 40 percent of the staff of USAID was returned PCVs (Hoffman 1998: 257). Of those who served between 1961 and 1995, 94 percent said that they would make the same decision to serve again.

CHAPTER 12

Doing Africa:
Travelers, Adventurers, and
American Conquest of Africa

Kathryn Mathers and Laura Hubbard

> In the adventure we forcibly pull the world into ourselves. Adventure has
> the gesture of the conqueror, the quick seizure of opportunity, regardless
> of whether the portion we carve out is harmonious or disharmonious with
> us, with the world, or with the relation between us and the world (Simmel
> 1971: 193).

Adventure, as Simmel suggests, is analogous to a love affair—it requires both
an act of conquest and a submission to the conquered. This simultaneous
conquest and surrender characterizes contemporary travel by Americans to
Africa. We argue that it is precisely the act of submission, the giving in to
love, that makes the enacted conquest of adventure palatable to the modern American subject. This not only intimates that travel to Africa actually
requires an "adventure," but that the ideal type of the adventurer is critical
to the imagined relationship Americans have with the world, one that is often played out against the backdrop of an African scene. The collapsing of
histories of representation of Africa as a colonized space of darkness and
contemporary images portraying a continent where modernity fails and
epidemics run rampant creates an Africa located fully outside of the United
States. America's Africa is the definitive site of adventure. An examination
of travel to Africa by Americans shows both a frantic yearning for adventure and a frequent reiteration of love and desire for Africa. This hunger for
adventure suggests that imperial relationships of the present are negotiated
increasingly through modes of action rather than the gaze. This shift from

the imperial eye/I to an embodied penetration of Africa by Americans pushes the American adventurer to realize a self more suited to emerging forms of empire. Drawing on ethnographic work with college students and other travelers to southern Africa, literature on African tourism, including sex tourism, as well as the role of Africa in reality television and other fictions, this paper shows how Americans forcibly pull the world into themselves.

We begin by investigating the ways "the gaze" continues to dominate writing on travel, suggesting that embodiment, performance, and action, while not displacing the gaze as central to tourism practices, are fundamental to grasping adventure's link to emerging imperial relationships. Through examining sex tourism in Africa, we illustrate these embodied erotics of adventure. To illuminate how Africa becomes the definitive site of the adventurer's encounter with the self, popular images and discourses about adventure travel are analyzed. Then, articulating the relationship between the distinctness of America's Africa with the neoliberal subject-making process, we discuss the widely circulated adventure of CBS Television's *Survivor Africa*. Here we move from a discussion of adventure's relationships of power in general to a specific center of power situated in the United States. Finally, we suggest how the actual practices and desires of American travelers to Africa demonstrate how "doing Africa" requires a romance narrative.

From the eye/I to the body:
the erotics and embodiment of adventure

The gaze is a fundamental trope in both historical and contemporary travel writing, not least because of its role as a primary western/modern form of appropriation (Pratt 1992; Spurr 1993; Urry 1990). Sight is privileged above all other senses so that it achieves an authority lacking from other forms of perception (Crawshaw and Urry 1997). The primacy of the visual in "encoding and legitimating economic expansion and empire" is clear in Pratt's "the seeing man" and "master-of-all-I-survey tropes," which were fundamental to the way travel writing emptied landscapes for imperial appropriation (1992). In contemporary tourist studies "the tourist gaze" persists as a powerful mechanism for writing about how places are structured at home and abroad for tourists.

The gaze was and is highly gendered as normatively male (Craik 1997; Ryan and Kinder 1996). "It is hard to think of a trope more decisively gendered than the monarch-of-all-I-survey scene. Explorer-man paints/possesses newly unveiled landscape-woman" (Pratt 1992: 213). Leed goes so far as to describe "travel as the unbridled freedom for masculine aggression and sexual adventure," a "spermatic journey" (Rojek and Urry 1997: 17). Jokinen and Veijola (1997) argue that just as it is the male imaginary that supports the western symbolic order, so figures like the tourist and the adventurer are masculine metaphors. Ebron shows how traveling and masculine agency are intertwined: "cultural stability is the imagined female to the travelling male" (Ebron 1997: 228). Such male desire and control was equated with love of the land. Imperial travel writing sexualized the landscape suggesting a male control of a female hysteric who is ultimately dominated but never entirely possessed. But just like a woman, Africa though dominated, can, according to Stanley, demasculate a European man (Spurr 1993). There is something ultimately feminine in even the male traveler to Africa (Clark 1999).

Eroticization, another key trope of imperial travel writing, required a phallocentric discourse as part of the colonialist discourse (Said 1979; Spurr 1993). This masculine penetration of interior spaces that asserted the colonial writers' mastery and the power of the empire, gets deeper in the contemporary days of sex tourism in which "men's bodies [are] associated with doing, action, and women's bodies [are] there for men's gaze or use" (Pettman 1997: 95). Tourism brochures and advertisements show how tourism privileges the male heterosexual gaze through creating images of masculinized northern adventure landscapes and feminized Third World landscapes. "Landscapes are shaped by the discourses of patriarchy and (hetero)sexuality and … the language of tourism promotion is scripted for a male heterosexual audience" (Pritchard and Morgan 2000: 886). Tourism images are infused with assumptions of masculinity as adventurous, action-oriented, and of femininity as based in domesticity and the opposite of "doing." Tourism advertising still assumes a male, white, and heterosexual tourist, thus continuing to privilege the gaze of the "master subject" (Pritchard and Morgan 2000: 889). Tourism continues to be studied as a staged authenticity that appeals to male fantasies. Such eroticization depends on the gaze. Even while suggesting that tourism is increasingly recognized as corporeal, Ryan and Hall argue that the gaze remains privileged even in sex tourism (Ryan and Hall 2001). Although ethnographic studies of travelers are increasingly breaking the mold by firmly situating tourist experiences as embodied (Frey 1998; Harrison 2003), and examining the implications of a "reverse gaze" (Sumich 2002), this emphasis on the gaze contributes to the continued absence of the body in tourism studies.

The history of women's travel writing and its analysis complicates this story. Initially disregarded as frivolous and trivial, women who undertook adventures during colonial times and wrote to tell about it, were initially portrayed as queers and eccentrics (Mills 1991). Pratt's reading of Mary Kingsley argues that instead of the master-of-all-I-survey sensibility, Kingsley uses humour and comic irreverence, and while supporting European expansion she does so within a stance against domination (Pratt 1992). The cross-dressing Isabelle Eberhardt traveling through North Africa as a Muslim man in the late nineteenth century is a complex, orientalist, and colonial figure. Her practices ran counter to many colonial ideologies (Smith 1995; Eberhardt 1991) as she pushed the boundaries of self and other, colonized and colonizer. Yet Eberhardt, while troubling most of the colonial binaries of race and gender, replicates the colonial narrative of self-discovery through detour of the "other" (Rice 1991). Travel by women can unsettle the continued dependence on the gaze as structuring metaphor for the analysis of both the adventures of the present and those located within colonial travel writing. A profound example of this is that while Ebron writes about the tourist gaze as "the medium for creating contradictory stories of a national agency," her work with Gambian men and the European women with whom they have sex suggests that sight is not what is primarily engaged in this structuring of national and transnational relationships (1997: 233). While women's narratives and bodies complicate Simmel's easy gendering of the adventurer, they remain complicit and within the love story of self encounter enabled by Africa.

Work on sex tourism challenges the hegemony of the gaze as it begins to propose ways of developing an embodied understanding of tourism and travel and suggests links between travelers and adventure. Much of the literature on sex tourism has tended to focus on the economy of sex tourism and the South East Asian industry, which is dominated by men seeking sex with women or boys (Pettman 1997; Ryan and Kinder 1996; Taylor 2001). More interesting here is the literature on women who travel for sex or have sex while traveling because such practices potentially undermine the masculinity of both travel and adventure. Often women who have sex while on vacation see themselves and are perceived as engaged in romance tourism rather than sex tourism (Herold, Garcia, and DeMoya 2001; Taylor 2001). This "Shirley Valentine syndrome" suggests that Western women can head south and have sex with Caribbean or African men in order to invest their own lives with meaning prohibited by the gender roles they are forced to play back home. Such analyses do not take into account the unequal power relations that exist between European and American women and Caribbean and African men (Ebron 1997; Pettman 1997).

Focus on the body in colonial writing was deliberately external, in fact assuring a suppression of the erotic. Travel is marked by an almost extraordinary denial of sexual congress that leads to the journey becoming "the search for and reconfirmation of the loss of love, and consequent redirection of an almost libidinal intensity upon the self" (Clark 1999: 21). This is not to suggest that sexual relations did not occur between colonizers and colonized, but that sex and sexual identity was absent from the writing that was used to assert power and dominance. Colonial officers referred to their African mistresses as "sleeping dictionaries" (a way for them to learn the language), implying that the book of Africa could be read through sex with an African woman (Spurr 1993). In the present, however, colonized bodies are sought out for sexual encounters. Sex for today's travelers is a mode of power, asserting their genuine experience of Africa, suggesting their acceptance of racial differences, their political and social liberation/liberalism, and their commitment to the "oneness" of the human race. The African body is still part of the landscape; it is just that now if you want to "do" the landscape you must "do" the bodies. Ebron quotes Mary, an English tourist: "You don't ask for sex, but the men seem to know you want it. They say things like, 'Would you like to see the real Africa?'" (Ebron 1997: 223).

Mary powerfully illustrates how "doing" an African is a requirement for experiencing "the real Africa." Ebron argues that tourism for sex between European women and Gambian men expresses important issues around national anxieties over power differences between Africa and Europe and between men and women. Women travelers are seldom referred to in complimentary fashion because women "can never carry national hopes and dreams in their travels" (Ebron 1997: 224). European women in Africa do not live according to the power relations they abide by at home. Their foreignness marks them as something other than women in their relationship with African men. They are emancipated from the male dominance of their ordinary lives. Sex for the women was part of their journeys of self-discovery. Their imagined Africa is "a place of free expression and sensuality ... that primordial place when one can transcend the confines of a constricted life in the North" (Ebron 1997: 240). Ebron refers to the Gambian men who seek a way out of Gambia and/or poverty through these women (using them to gain entry into European countries, possibly for marriage and ultimately work and residency permits) as "participants in transnational tourist adventures" (Ebron 1997: 225).

There is certainly an element of freedom in being removed from the standard structures of control that prevail in everyday life. Many people are different when they travel. Yet the tendency in most international travel is that it is dominated by structures that make people feel at home. Adventure travelers may put their lives in real danger, yet they maintain a lifestyle that looks very much like the one they left. There is minimal attempt to get to know strangers, to risk one's emotional and mental comfort zones and disturb any preconceived ideas about the places they are roaring past (Vivanco 2001). The reality programs that are sending people all over the world in fact require very little adventure or excitement to be derived from the actual places that the participants pass through. Similarly, the way sex tourists talk about their use of prostitutes while traveling suggests that such activity is defined by the paradox of the marginal act, which is that it is essentially a conventional act undertaken for conventional reasons (Ryan and Kinder 1996: 515). Observation of tourists who engage in adventure suggests that they are often simply conforming to the requirements of contemporary travel.

Doing Africa, Loving Africa:
Africa as distinct site for Simmel's adventurer

Here, however, we are more interested in the national anxieties between Africa (although a continent) and the United States that are played out through adventure travel and adventurous travel. If sex tourism is "a classic moment in international relations [as] pleasure and danger come together with transgressions across borders of power along First World–Third World, Rich-Poor, male-female (often), old-young (often) in a peculiar and unstable combination of sexuality, nationalism and economic power" (Pettman 1997: 101), what is adventure tourism, which mimics sex tourism in its penetration and its essentially male action? Here we explore the possibility that the phallic penetration of Africa occurs boldly/bodily through "doing Africa" and that it is achieved equally by male and female adventurers.

Adventure travel is one of the fastest growing arenas of the tourism industry, especially among North Americans and women (Neirotti 2003). It is considered mostly in terms of its role in the broader travel industry and the challenges inherent in trying to define something like adventure, let alone adventure tourism (Swarbrooke et al. 2003). Cloke and Perkins's analysis of the way the increasingly popular New Zealand adventure travel industry is changing the configuration of New Zealand's landscape tells the story of how adventure can tame a landscape (Cloke and Perkins 1998). The connection between "paradisiacal nature and adventurous activity, in which tourists are encouraged both to gaze at spectacular scenery and grapple with the challenge of nature" is a fundamental part of adventure tourism (Perkins and Thorns 2001: 196). Adventure travel augments the tourist's gaze in ways that require paying more attention to what tourists feel, touch, and do (Cloke and Perkins 1998). Performance may more adequately capture the experience of adventure tourism than gaze because "it connotes both a sense of seeing and an association with the active body, heightened sensory experience, risk, vulnerability, passion, pleasure, mastery and/or failure" (Perkins and Thorns 2001: 196).

Travel brochures on Africa available in the San Francisco Bay Area tend to be largely confined to high-end, extremely luxurious safaris or down-end adventure tours. Adventure travel specialists have increasingly been adding Africa and southern Africa to their itineraries. Even Contiki, the famous provider of bus tours around the globe to eighteen to thirty-five-year-olds, offers overland safaris to South Africa, Botswana, Namibia, Zambia, Zimbabwe, Kenya, Tanzania, Zanzibar, and Swaziland. Such tours use overland vehicles and emphasize opportunities for "adventure," such as mountain biking, bungee jumping, and white-water rafting. Turning to an examination of two such advertisements, the distinct nature of Africa as a site for adventure takes center stage.

Opening the brochure for G.A.P, "the great adventure people," not to be confused with the retail clothing chain of the same name, a sparkling lake surrounded by craggy peaks quietly greets the reader with one word in bold inscribed in the sky: Earth (G.A.P 2003/2004). The introduction on the following page discusses how the adventurer must step outside of controlled travel in order to watch elephants in Botswana or sit with gorillas in Uganda. The overleaf includes a recommendation for G.A.P written by Don George of Lonely Planet publications, the proponent of adventure through guidebooks worldwide. Describing his lifetime of adventure, George recounts giving a postcard of San Francisco's Golden Gate Bridge to a family with whom he spent the day, and comments that this interaction is "the kind of connection that the adventures described in this catalogue can bestow— encounters and lessons that transform us." George continues, "the good news embodied in these pages is that, as with my ascent of Kilimanjaro, you can make those dreams come true…. Within these pages, Africa offers "everything an adventurer could ask for," emphasizing dune boarding, boating, and the climbing of things. In the end, "Africa will touch you deeply and beckon you to return again and again" after visiting the "undiscovered game parks of Zambia" and reveling on the summit of Kilimanjaro. Lower budget "overlander" companies like Exodus rely almost exclusively on offering up a *real* Africa for a *real* adventurer (Exodus 2003/2004). For the consumers of Exodus, crossing Africa is a "true adventure" and indeed, Southern Africa is "one of the most exciting areas in the world to discover." The African overland adventures are described as life-changing, and after white-water rafting, bungee jumping, kayaking, diving, hiking, and sand boarding you will apparently finally understand that "you can leave Africa, but Africa will never leave you." Photos in the overlander pamphlet show travelers penetrating the desert with the truck, celebrating their summit of peaks, canoeing and rafting the Zambezi—they are truly "doing Africa," an activity that intoxicates them with erotic love. Africa is the backdrop for the encounter with the self through adventure, marketed as the dream come true for listless Americans.

"Victoria Falls is just another waterfall": Plunging into Africa

The absence of tourists in much of the writing on travel and tourism, even when the stated goal of the work is to build a "performative approach to tourism," has hampered our understanding of tourist experiences (Colls 2003). One of our projects set out to use ethnographic research to understand the impact of travel to South Africa on American attitudes to Africa (Mathers 2003). Here we draw on parts of this ethnography to show how American travelers, especially young ones, tend to experience Africa through actions, and in fact, require adventure in order to feel that they have been to Africa. This work consisted of three years interviewing and observing Americans who had or were planning to travel to South Africa and elsewhere in

Africa, as well as five months participant observation of study-abroad students from California in Cape Town. It was with this latter group that the search for adventure coincided with the search for Africa.

Despite the multitude of warnings about the physical dangers they would face in Africa, these young Americans were convinced that they had not been to "Africa" until they had bungee jumped at Victoria Falls or done the white waters of the Zambezi on an inner tube. Gazing on the Victoria Falls itself was considered completely unnecessary for experiencing Africa. American study abroad students to Cape Town focused much of their energy on seeking out opportunities for the best skydiving, bungee jumping, white-water rafting, and boogie boarding on the Zambezi that they could afford. The short-term trips that they took for their midterm break took advantage of package tours clearly offered specifically to attract them. Although one group of students drove up the coast from Cape Town, most flew or caught a bus to Johannesburg and joined a tour going to Kruger Park and then on to Victoria Falls. The longer trips headed straight north to Botswana and Zimbabwe. Even the purpose of driving up the coast was to bungee jump at Tzitzikama (the highest in the world) before heading to Victoria Falls via Kruger Park. One student had bought a big hunting knife for the safari and many of the students were really surprised that you could not get out of your car in a game park.[1] They all talked as if they were going to the jungle but at the same time not really believing that it might be truly dangerous. Victoria Falls was a must despite the fact that no one seemed to know which country it was in (Zimbabwe and Zambia). One study abroad student, extolling the virtues of white-water rafting, actually referred to Victoria Falls as just another big waterfall, "it's the Zambezi that's hot."

South Africa seems to stand in contrast to the north/adventure, south/feminine passivity dichotomy in that it is central to an adventure travel industry that dominates the bottom end of its tourist market. The landscape's gendered configuration has not changed. Africa remains figured as female, inviting yet wild and dangerous. But it is no longer conquered with a gaze, and it must be physically penetrated. South Africa may offer the "feminine seduction and masculine adventure" (Pritchard and Morgan 2000) that is supposed to appeal to a largely male, heterosexual gaze but female action is penetrating these landscapes. It seems ironic that a location like southern Africa should be so successful at building a market for such adventure tourism. Few people can plan a trip to this part of the world without being constantly warned of the dangers both at the hands of people with guns and no scruples and Africa's multiple pathologies. Disease and death keep parents awake at night worrying about their children in this part of the world. It is particularly intriguing then that they should, once there, choose to engage in dangerous behavior. Often contemporary adventures in Africa are marked on the travelers' bodies physically through piercings that they would never have had the courage to get at home. This danger creates the intimate and embodied love affair with Africa that is consummated by the returned traveler's determination to seek succor for the ills of this continent.

Colonial travel was characterized by the reduction of the nonwestern world to an inner journey, and the importance of personal change remains one of the key motivators for all forms of travel (Hastings 1988; Spurr 1993). Adventure travel is certainly motivated by a similar need to become a better person, not just internally but physically as well (Rojek and Urry 1997). This latter component marks a fundamental difference between past and present as the physical impact makes it harder to reduce adventure travel simply to an inner journey. Adventure tourism does the opposite of observing from a distance; contamination is, in fact, exactly what adventure travelers are looking for. But the relationship between imperial power and periphery remains the same.

Photographs have always given shape to modern travel and tourism. They appropriate, transcribe, aestheticize, miniaturize, semioticize, and democratize. Tourists are often just tracking down images they have already seen (Crawshaw and Urry 1997). In adventure travel the photograph is now almost invariably a moving image, and it is a moving image of the tourists themselves, not of the object/space/place being penetrated. When travelers were interviewed after their return to the states, their photographs gave a good sense of the events and interactions from their time in South Africa that they considered important. Young Americans who leave southern Africa are more likely to have a photograph (a still from the video) of themselves bungee jumping or kayaking at Victoria Falls than an image of the falls. The same can be said of the new generation of wildlife programs that are more about the person/showman interacting with the crocodile or snake than they are about the crocodile or the snake. *Survivor Africa*, for example, is a visual testimony to the traveler's experience, not to the game park in Kenya. These travelers who penetrate the very space that is normally overseen and compared with home disturb the trope of the commanding view, so common in imperial travel writing.

This all speaks to a fundamental shift in the way travel represents/ constructs the relationship between the periphery and the metropolis and also possibly the differences between American and European imperialism. Travel is no longer a search for familiar images but is driven by the need to repeat an experience. Travelers want to get into the bodies of the travelers before them, not stand in the spot in which they stood. Importantly, these bodies can be male or female. Regardless of whether one reads the iconography of adventure travel as gendered or masculinized, young travelers enact it androgynously. This is particularly ironic because much of the action, the doing of adventure tourism is singularly phallic and penetrating. Like Tarzan, these travelers do not "see" Africa, rather they "do" Africa. This doing Africa by plunging into "her" center underlines the extent to which Africa has not only been gendered by this new generation of androgynous travelers, but sexualized and penetrated.

Encounter with the neoliberal self: American "Reality" in Africa

Survivor[2] is a reality television show that drops sixteen Americans onto emptied-out landscapes in far corners of the world, dividing them into tribes, and with twenty-four-hour surveillance and fantastic editing, allows audiences across the world to watch as the Americans form alliances that they must eventually destroy in order to emerge as the sole survivor. *Survivor Africa* and its drama, which played out in the Shaba game reserve in Kenya, demonstrated that colonial modes of representation and knowledge coexist with more contemporary forms of empire and the neoliberal humanitarian world order. This was exposed through the relationship the contestants formed with the land, through the encounter they had with themselves and the moments of submission to Africa portrayed within the show. Cut off from the world with absolutely no contact with "the outside," they are offered various landscapes to conquer—it is hard to imagine a more perfect televised version of Simmel's adventurer than the embodied experiences of the *Survivor* cast members. Shrinking and shriveling in the harsh elements, the contestants struggle to overcome the land and each other. *Survivor* proposes an adventure for millennial forms of empire—unsexed, with a receding but still important love story.

Through its very naming and the careful manipulation of land and people, *Survivor Africa* allows an erotic love story to emerge between the contestants and "Africa." This is no trip to the tropics, like most *Survivors,* and
it shows, in both the encounter the participants have with themselves and
the love story that emerges. On this adventure, and not on *Survivor Thailand,* or *Survivor Amazon,* the Americans land in a place devoid of specific
geography, they arrive in "Africa." The parameters of this action of naming
illustrate that there is one place that can still be considered as without history, and that is Africa. This allows the show to sample from periods of history and sites without concern, citing both nostalgia for colonial Africa and
the modern AIDS epidemic simultaneously. The very naming of the show
lays the foundation for a devotion to the land to emerge among the contestants; a love affair flourishes with unspecified "Africa" lending them a
backdrop for the discovery of themselves.

The second element of *Survivor Africa* that is radically different from
all Survivors past and present is the representation of "modernity" within
the show—a hospital, AIDS, and a town intrude into the landscape, which
in other *Survivors* is usually emptied of natives. That the trappings of the
modern world can appear in such a show and not disturb the remoteness
story so central to its success underlines for every aspiring neoliberal subject that African's modernity is a failed one. This degenerate modernity and
its attending plague become the ground for the love affair of the adventurers. It is particularly important to point out that the effusive love the *Survivor* contestants share for Africa is not an aspect of any other *Survivor*
shows past or present, whether in Polynesia, Australia, Thailand, or the
Amazon. In this *Survivor* and no other, contestants fall in love while coming face-to-face with the self most suited to carry on America's empire. This
could only happen in America's Africa.

Without any geography, without specificity, the inhabitants of *Survivor
Africa* move in a passionate relationship with the land. In no other *Survivor*
do participants speak in such tongues. One woman continuously talks about
"mother Africa" the love due to her, as it is this mother Africa that gave
them life. Frank, the military man, declares in a confessional interview dis-
cussing his desire to be one with his surroundings that "To be here is just
awesome. My love for being here just grows every day (episode ten, *Fare-
well to Frank*). Tom the goat farmer and Lex the musician win a challenge
with a reward to do Africa by floating over the vastness of the plains on a

hot air balloon safari, talking incessantly about the awe-inspiring sweeping plains (Episode eleven, *Survivor Safari*). These *Survivors'* love affair with the land is encouraged to reach the height of its imperial longing when Brandon, the gay bartender, and Frank win a reward challenge to go to the movies (Episode nine, *Out of Africa*). Watching *Out of Africa*, Brandon and Frank are primed for a lesson in submission to love—to give themselves over to the land, a place imagined for them in the film as one to care for tenderly, but empty of such cities as Nairobi. In no other *Survivor* is a reward challenge staged in which the participants view a nostalgic old film, or a film of any kind. The show itself teaches them to court the very land, to foster a passion for mastering it. In no other *Survivor* do the participants discuss return—these survivors spill over with a love of Africa, a love instilled through colonial images, a love that denies the presence of Africans, a love that functions very well as the foundation for a kind of caring appropriate to contemporary incarnations of empire. No show is a map for the contemporary adventurer, and the neoliberal subject, like *Survivor Africa*. *Survivor* posits the adventurer as ideal American citizen of a changed empire. It offers a primer for how Americans can pull the world forcibly into themselves through a story of love and an act of conquest.

Love and Sex: Reciprocity between Africa and America

> A love affair contains in clear association the two elements which the form of the adventure characteristically conjoins: conquering force and unexttortable concession, winning by one's own abilities and dependence on the luck which something incalculable outside of ourselves bestows on us (Simmel 1971: 195).

A tension between power and love is found in past imperial discourses, especially that of the anti-conquest or "the strategies of representation whereby European bourgeois subjects seek to secure their innocence in the same moment as they assert European hegemony" (Pratt 1992: 7). The anti-conquest is prevalent in the writing of the "sentimental hero," Mungo Park, whose constant exposure to danger and to the ridicule of the native both human and geographic, suggests that he is no conqueror (Pratt 1992). Park's struggles with natives and other creatures suggest his attempts at achieving a reciprocal relationship with them. This requirement of reciprocity is a classic

myth of capitalism (Pratt 1992). The link between capital and past and present imperialism makes it, therefore, unsurprising that we should see in the
language of contemporary American travelers to Africa an assertion of what
they are giving or will give to the continent. They are in their turn creating
their own language of the anti-conquest, but it is a language that does not
depend on the gaze as much as it does on action.

Pratt suggests that romantic love was an expression of cultural harmony between colonizers and colonized. Reciprocity is required to fuel the
illusion that some sort of equitable relationship existed, and requited love
embodies this ideal as "romantic love like capitalist economy understands
itself as reciprocal" (Pratt 1992: 97). *Survivor Africa* and adventure travel
to southern Africa play out this illusion of reciprocity that is so fundamental
to the illusion that the U.S. is not an imperial power in relation to Africa.

Travel to Africa for American students plays out like a love affair. First
anticipation, excitement at what lies ahead with plenty of unrealistic perceptions of what it might be like. Then a honeymoon period where everything is different, wonderful, and new, but once that first blush of wonder at
its beauty and complexity is over, there is disillusionment. South Africa is
not quite what it should be, it is scarier, politically—complicated, violent,
cynical, and not that into them. Lastly, on return, having abandoned the
place, they look back with love and a sense of responsibility, a need to return so South Africa knows they only left because they had to, or to devise
ways of "helping." The travelers' journal entries illustrate this trajectory:

> At the end of my first day, I already realize, there is no way I am ever going
> to be able to tell anyone about this trip. I can tell people what I have done,
> but that will not tell the story. It is the feelings I am having that make up
> the essence of this adventure, and there are simply no words that I can de-
> scribe my feelings. When people ask me "How was South Africa?" I hes-
> itate before answering, because it is impossible to synthesize all of my
> experiences and impressions into a tidy, compact sound bite (Johanna).
>
> Has South Africa changed me? I told him that while I was there I was
> in many ways a very different person. I felt more carefree than I have in a
> while. Each day was fascinating and exciting and so different from the day
> before. Each day I felt we experienced the one thing I would hold onto as
> the most memorable part of the trip and then the next day we would do
> something else that I was convinced would be the most memorable. That
> sort of existence changes a person. It makes life seem so exciting and in-
> teresting. And above all, everything we did was imbued with a sense of

urgency. Perhaps being in a country where incredible amounts of change have occurred so recently creates that sense of urgency (Teresa).

I'm going to apply for positions with USAID in hopes of getting sent to Africa. I want to pursue a career that will allow me to travel to Africa and possibly influence issues that affect Africans (Caroline).

Survivor Africa also demonstrates the critical way the travelers encounter selves that are the result of their relationship to an imagined Africa, one that requires reciprocity. Lex, a rocker from Santa Cruz, California, wins a challenge that garners him both a Chevy Avalanche SUV hybrid and the "opportunity" to deliver AIDS testing kits to a rural hospital (Episode twelve, *Wamba Hospital*). The fact that the modern epidemic and a hospital intrude into *Survivor Africa* is not surprising, though no markers of modernity have ever trod on the pristine supposed wildness of the other *Survivor* sites. Africa's modernity is so emaciated that instead of bringing it closer to America, it underscores Africa's remoteness. Coping with this failure of modernity becomes the romance of *Survivor Africa*—the need to help and to care emerge as the new erotics of the NGO world order. Lex and his comrades submit not just to mama Africa's wide-open grasslands, but to a desire to doctor the continent's ills. As the show draws to a close, most of the participants have given in to this particular form of love, the care for the African afflicted with AIDS. Brandon, the gay bartender, holds a fund-raiser, and all the contestants speak of giving back and of return. The sympathy engendered by failed expectations for transformation gives rise to a form of care that rides on the belief that America can save Africa, that within the realm of responsibility for freedom is the need to care about Africa. This burden of care serves the "just" notions of humanitarian aid and the regime of the NGO and operates to further the narrative of self-discovery for the contestants. Africa is penetrated, conquered, emptied, mastered, and finally, loved through the adventure of discovering, not Africa, but the self appropriate for this world, a subject whose ideal character is found in Simmel's adventurer.

Penetrating conclusions

The adventure lacks that reciprocal interpenetration with adjacent parts of life which constitutes life-as-a-whole. For this reason, the adventurer is also the extreme example of the ahistorical individual, of the man who lives in the present (Simmel 1971: 189–190).

The likelihood that both the "Africa," "America," and the "adventure" de-scribed here are imagined (like the love affair) and that these adventures do not require a male subject does not rob us of the analytical power of Sim-mel's ideal adventurer. In particular this adventure that is "certainly part of [their] existence, directly contiguous with other parts which precede and follow it: at the same time, however, in its deeper meaning … occurs out-side the usual continuity of life" seems ideally suited to creating the model American citizen in relation to spaces like Africa.

The adventurer as an ideal type is one of the richest and most produc-tive exemplars developed by Simmel for the analysis of the making of mod-ern selves and worlds (Frisby and Featherstone 1998). By investigating both the actual practices of travelers and the widely circulated images of jour-neys to Africa, we show that through utilizing Simmel's analytic category in tandem with thinking about the neoliberal subject-making process, travel to Africa becomes the ultimate adventure and poses the model citi-zen for a liberal America. We have used the "adventurer" to consider what is constituted as the gesture of the conqueror and its attending act of love in the contemporary, arguing that love of the land and submission to its emptied vistas is no longer enough—both an ethic of care and a relocation of experience from the gaze to the body are demanded of the neoliberal ad-venturer (Bourdieu 1998). Bourdieu asserts the pivotal cultural work the adventure plays in neoliberalism by describing its subjects as both "schooled in higher mathematics and bungee jumping."[3] The bungee jumping esca-pade at Victoria Falls, so far removed from the boardroom and cut off from the ringing of cell phones, for Bourdieu, is the embodied plunge into the self most adept at managing the stock trading floor.

It is no accident that Africa is the place where the adventure is recon-figured and recirculated in this way—it is, we argue, the definitive site. From narratives of travel to Africa to its televised versions such as *Survivor Africa*, *MTV Adventure Cribs in Cape Town*, and even Oprah's Christmas trip to South Africa in 2002, it becomes clear that Africa is still the training ground for citizens of empire, this time one whose currency is the discourse of the NGO and the free market. The world is forcibly pulled into the American through a language of love for Africa, one that masks a fundamental ges-ture of conquest. What Simmel offers explicitly that other theorists of this subject-making process do not is the embodied erotics of the adventure, demonstrating how tightly colonial forms of power and travel are bound with contemporary modes of empire, action, and subject formations. Simmel poses that the adventurer is essentially a male figure submitting to and yet conquering a feminized form. While women come to inhabit the adventurer's body more often, the gendered nature of the power relations embedded in the erotics of the adventure is part and parcel of its current territory, regardless of the sexing of the subject themselves. Through the use of the adventurer as ideal type, the domain of travel of Americans to Africa becomes essential to understanding power.

> That one has faced mortal danger or conquered a woman for a short span of happiness; that unknown factors with which one has waged a gamble have brought surprising gain or loss; that physically or psychologically disguised, one has ventured into spheres of life from which one returns home as if from a strange world—none of these are necessarily adventure. They become adventure only by virtue of a certain experiential tension whereby their substance is realized. Only when a stream flowing between the minutest externalities of life and the central source of strength drags them into itself; when the peculiar color, ardor, and rhythm of the life-process become decisive and, as it were, transform its substance—only then does an event change from mere experience to adventure (Simmel 1971: 197–98).

This empire centered in North America requires the submission that comes with the fear and the danger associated with adventure travel. A romance narrative emerges that performs a conquest through its very telling: that good American citizens must submit to a love for Africa in order to support the agendas of the NGOs and other agencies as they broker peace and bring AZT to the continent, while not even pausing to consider the gesture of empire that the nature of their adventure in love enacts.

Notes

1. These discussions were largely amongst a group of semester abroad students from a small private college in northern California. They spent five months in Cape Town, studying at the University of Cape Town but under the guidance of a faculty member from their college. The seventeen students all lived together in a large house near the campus. They were diverse in terms of race, ethnicity, class, and gender.

2. http://www.cbs.com/primetime/survivor3/

3. http://mondediplo.com/1998/12/08bourdieu

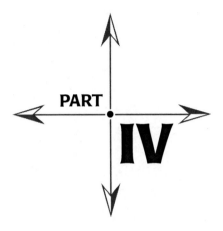

PART

IV

Cross-Cultural Adventures

CHAPTER 13

"Oh Shucks, Here Comes UNTAG!": Peacekeeping as Adventure in Namibia

Robert J. Gordon

The ending of conflict is a specific enterprise. It belongs neither to war nor to peace, just as a bridge is different from either bank it connects (Simmel 1955: 110)

Rereading Simmel, one is struck by how his view of the Adventure creating an "exclave" is similar to Victor Turner's notions of anti-structure, liminality, and communitas. For Simmel the Adventure is largely experiential fantasy, others and ours, packaged as time away from ordinary life. Most studies of adventure have focused on individual adventures, and this paper suggests that one can extend the notion by incorporating the Turnerian dimension. Transitions, as Turner showed, are situations of uncertainty and danger in which one moves to a new status. Conventional analyses of Adventures see them as entailing a distinctive structural movement, either spatially and/or socially, in which the adventurer has to deal with a challenging element of unpredictability, risk, and danger, all of which combine to create a "Time Out" from the operation of the daily humdrum of society (with the challenging element of danger).

A better sense of what is meant can be gained from looking at the roots of the word. Adventure is derived from the Latin *ad*, meaning "toward" and *venire*, meaning "coming." An adventure then has that tingling air of expectancy of something about to happen. This tingling is also coupled to an

awareness of danger or fear, no matter how dominant or niggling. Camus
(1962: 26) said it well:

> What gives value to travel is fear. It breaks down a kind of internal struc-
> ture ... stripped of all our crutches, deprived of our masks ... we are com-
> pletely on the surface of ourselves ... this is the most obvious benefit of
> travel.

It was this aura of expectancy that is intrinsic to the experience of adventure that was so blatant among the personnel of the United Nations
Transitional Assistance Group (UNTAG) who were overseeing the transition of the de facto South African colony of South West Africa to the independent country of Namibia. Given that UNTAG literally and figuratively
assisted in the change of status of South West Africa to Namibia, the participants clearly served as ritual elements designed to facilitate this transformation. Adventure is very much a staged performance (MacCannell
1989). The ways adventures are performed do not merely reflect views of
reality but create and confirm them.

In 1989 after about a decade of international negotiation against a background of constant low-intensity guerrilla warfare against the SADF (South
African Defence Force) by SWAPO (South West African People's Party), the
first U.N. "multi-functional peacekeeping" exercise took place when about
nine thousand members, mostly troops, of UNTAG arrived in Namibia, one
of the world's most sparsely populated countries, with a population of about
1.5 million, to keep the Peace and to organize "Free and Fair Elections" as
part of Namibia's transition from disputed territory to an independent
democratic state. In undertaking this task it faced many challenges, including resettling some forty-one thousand refugees, organizing elections
where illiteracy was estimated by the U.N. to range from 60 to 80 percent
(Griffith 1998: 157), removing all discriminatory legislation, and monitoring the police in an effort to prevent harassment, especially by police "ir-
regulars" (known locally as *Koevoet* or Crowbar in English). During this
transition I was recruited by the United Nations as a consultant to make
recommendations concerning how the Apartheid Second Tier Authorities
should be replaced. It is on these experiences, coupled to a reading of UN-
TAG memoirs and reminiscences, that this paper is based.

Most UNTAG personnel saw themselves, and were seen by Namibians,
as having an adventure. Could it be that how they defined and acted upon
the situation contributed to making UNTAG the success it was proclaimed

to be? This adventurous attitude, indeed it resembled a temporary *weltan-schauung,* was in sharp contrast to the departing South African military. Like UNTAG, the SADF troops were largely male, younger, and uniformed, and had traveled away from home for their "tour of duty," yet clearly except for a few individuals, the majority did not see their sojourn in Namibia as an adventure despite the fact that structurally it was a situation containing unknown dangers and risks and an eventual return "home." Given structural similarities between them, it is important to ask, why the UNTAG experience was defined as an adventure and a "success," while the SADF engagement was seen as neither adventure nor a success. How does one account for this difference and is there any relationship between "adventure" and "success" in peacekeeping? This is no idle question. Indeed, DRPs (Demobilization, Re-integration Programs) are now one of the "hot areas" of development practice by the World Bank and other donor agencies. One of the cases frequently invoked as a success of such transitions is that of Namibia (Colletta 1996; Goulding 2002; Hearn 1999; Griffith 1998). Recently Howard (2002) has argued that the critical factor was not so much the consent of the warring parties and the strong Security Council interests as that UNTAG was able to adapt to the needs of the postwar environment. How exactly were they adaptable and might not this be part of the adventure?

The South African Defence Force in Namibia

During the 1980s the SADF deployed some seven thousand troops, mostly conscripts, in the more populated north in what was known as the "Operational Area." This is the area where the South African regime had been fighting a low-intensity war against SWAPO guerrillas who were seeking independence for Namibia. From a White (South African) point of view it was a frontier, not only culturally and symbolically but a state perimeter area as well since it skirted the Namibian-Angolan frontier,[1] the border between what was then seen as Black Africa and White Africa. It clearly had the characteristics of a liminal situation. As one white soldier said:

> The guys were saying that it was freedom up there you could get away from the basics of disciplined structures of the force at the time, you could go up there, there was freedom, there was more money, they were having a good time.

Such a situation was promoted by an overwhelming sense that people in the "States" (South Africa) "just didn't care, they didn't know what we were really doing" "up there." (Verwoerd 2002: 23–24).

In 1985 a feisty little South African magazine appropriately named *Frontline* decided to do something about this ignorance and published a story, "Ends of the Rifle," which dealt with the border experiences of a young white South African national serviceman. The tenor of the story is clear:

> When I was up there I felt nothing.... You look at terrs [insurgents] as stupid. He doesn't know what he's doing. He doesn't know what's right and wrong.... You don't think of him as a human being.... They go to extremes. That's stupid. You can't see them as people.... There was a lot of rape. We had some low guys with us.
>
> Some of those guys you couldn't understand.... Also, when you're there you're *sommer* (just) fighting, you aren't thinking about what you're fighting about.... Mainly the *ous* (chaps) don't bother about prisoners. If you take a prisoner you have to go through a whole procedure, filling in forms and that. So you say he was shot in contact. Maybe you don't say anything. (*Frontline* 1985: 54).

In this geographical expanse the normal rules of society (and "civilized warfare") did not apply. Both prominent politicians and lowly soldiers publicly claimed that if they were attacked "no rules apply at all.... We will use all means at our disposal, whatever they may be" (*Weekly Mail*, 2 Sept. 1988). In this zone a new lexicon came into play based, rather intriguingly, on a vocabulary developed by Americans in Vietnam. This zone was populated by people who are the obverse of normal society. Recognizing SWAPO was easy, said our *Frontline* troopie: "If you see a big, healthy looking, male you know he's a Swapo. Only the Swapos eat well enough to look like that— apart from the Ministers ... and they're Swapos anyway ... If he had soft feet that would prove it beyond doubt...." (Anon 1985: 54). Similarly, while white women are seen to be nonactive combatants, loyal, upright, moral, and supportive societal exemplars, black women in the combat zone were the antithesis. The official Ethnology Manual of the South African Defence Force gave the following advice to young servicemen:

> Direct contact with Bantu women should be avoided, while <u>social contact should be ABSOLUTELY FORBIDDEN AND PREVENTED.</u> The comings and going of the Bantu woman should, however, be closely observed for

the following reasons: 1: She provides food and is the "beast of burden". It will most probably be her duty to provide the insurgent camp with food and beer ... and she may even carry the heavier automatic weapons, as happened in Angola. 2: She incites the men to successful military action by arousing them sexually before a battle, taking part in the "Washing of the Spear", a ceremony based on a party where sexual intercourse takes place. 3: Experience has shown that the Bantu will take advantage of the "weakness" of the Whites not to treat women roughly by pushing them to the front in political riots ... (SADF 1977).

In this scorpion dance the chief protagonists were not of this world. Our *Frontline* troopie claimed: "terrs are all on drugs—morphine and heroin (That's one way you can recognize a Swapo. You see the needle marks in his arms)," but goes on to admit that "some *ous* [guys] would get morphine from the medics, but most of us were just on dagga [marijuana]" (Anon 1985: 54). Even experienced pundits attributed extrahuman animallike qualities to the enemy. As one explained: "some of these terrorists are able to walk two or three kilometers at a stretch *on their toes* in order to confuse security forces. The imprint left in the sand is almost indistinguishable from animal tracks" (Venter 1983: 51). Nothing could be taken for granted in this theater, not even the audience.[2] A South African officer commented: "The Ovambo tribesman is very stoic and fatalistic. Simple killing does not have the impact on tribal members as does, say, cutting off arms and legs or noses and ears and then forcing the victim's wife to fry and eat the flesh" (Brown 1986: 53).

And then there was *Koevoet* or Crowbar, auxiliary police recruited from local people or "turned" SWAPOs, who lived a life of their own on the perimeter of the social order and who operated almost by their own laws. They were paid large bounties for every SWAPO terrorist they "brought in Dead or Alive" and would wear T-shirts proclaiming "Killing is our Business, and Business is Good." *Koevoet* was generally despised by both the local population and the SADF troopies. Plainly, except for a small minority who thought they were fighting to preserve the West from the onslaughts of Communism and black domination, most South African soldiers felt that that they were there against their will. Moreover, reading the numerous accounts and memoirs of service in the operational area it is clear that despite the Defence Force's WHAM (Winning Hearts and Minds) policy, the SADF was generally not welcomed by the local population.

The Blue Berets

UNTAG too were distinctive. The military wore blue berets and scarves while civilians had blue U.N. armbands. Their vehicles were all painted white and featured "UN" prominently painted on the side. They also clearly sensed an element of danger, which was reinforced at the start of UNTAG operations when on the day before the cease-fire was to be implemented the SADF engaged SWAPO in a bloodbath, alleging that SWAPO was trying to move troops into a strategic position prior to the 1 April cease-fire. By 5 April, more than 260 SWAPO combatants and 30 security forces had died in the heaviest fighting of the twenty-three-year-long war (Lush 1993: 147). This nearly scuttled the mission. One UNTAG officer, Col. Mwaneria, of the Kenya Battalion, or Kenbatt, described the Kenyan reaction as they prepared to embark for Namibia:

> It was not a reaction precipitated by fear, but the unpredictability of plunging into the unknown, the first step in a make or break situation. It may last a few seconds at most, then the normal flamboyant and easygoing manner wills itself over the doubts lingering in the soldiers' minds, and all is over. Only then is such a soldier ready to fight and fight hard, for, though death hangs as a possibility in the periphery of his subconscious it loses its sharp fangs.... (Once) reconciled to the new situation the troops jested on how they would go about their UNTAG duties. They saw themselves as acting as buffer to the flying bullets and missiles between the warring SWAPO and South Africans by standing in between and stopping the flying missiles with their own bodies. They laughed at the incredulity of the whole idea (Mwaneria 1999: 46).

UNTAG military activities consisted largely of patrols, flag marches, and participating in the many national day celebrations of the numerous countries participating in UNTAG. Kenbatt, for example, was constantly called to parade for visiting foreign dignitaries and celebrated two national holidays, *Madaraka* and *Jamhuri,* that attracted many spectators and guests and featured "traditional" dancing, sports, and choirs as well as the inevitable parade. Civilian UNTAGs, backed up by a slick multimedia campaign, engaged in a constant flurry of voter education that entailed visiting churches, schools, and numerous public civic events.

UNTAG, in contrast to the SADF, was generally welcomed in the Operational Area. Indeed, some observers felt that they were treated as "The

Liberators." It was in the south, the area where most whites and blacks opposed to SWAPO resided, that the UNTAG peacekeepers picked up some mute and occasionally active resistance, but this was largely attributed to the fact that they were blacks, in contrast to the Malaysian and Finnish troops stationed in the North. Indeed, black UNTAG members commonly feared that they would be subject to Apartheid, and in cases where this did surface UNTAG swiftly took redressive action.

Reactionary whites constantly accused Kenbatt of being in cahoots with SWAPO. Their chronicler believed that hostility from sections of the Namibian community was endemic. Stories of black UNTAGs being refused service at country hotels were common. Accusations and other intimidating mechanisms had become the order of the day. For the Kenyans, "danger was always lurking in the veld, in houses and along lonely streets" (Mwaneria 1999: 68) and reached its climax with the bombings of the UNTAG offices in Outjo, a small white town.

Most UNTAG personnel were volunteers, and apart from the high salaries such secondment brought, they were drawn to their assignment by a sense of history, duty, and adventure. In the years preceding the transition, South Africa's policy of *Apartheid* had been a constant idée fixe at the United Nations especially among the Afro-Asian bloc. Many of the volunteers from Third World countries stressed this, especially those coming from the Caribbean, who saw this as an opportunity to return to the ancestral continent and assist in the end of *Apartheid*. They also believed that the eyes of the world were upon them (the presence of an extremely large press corps emphasized this) and made them feel that they were participating in an important historical event. This notion is transparent in their accounts, memoirs, and even the kitschy doggerel published in the UNTAG house journal.

This extensive media coverage that preceded the operation meant that most UNTAG personnel had strong preconceived ideas about what to expect, namely an oppressed black majority, and most accounts continued to present simplified versions of the present and past, namely that local blacks were still badly discriminated against in terms of salaries and living conditions and that six o'clock curfews still pertained for blacks in the urban areas. Urban legends of white settler atrocities were also common (Mwarania 1999: 76). Similarly, in the Swiss UNTAG contingent's reminiscences, the standard clichés about Apartheid persisted in their descriptions of the social landscape, this, despite being chosen because of their expert knowledge of Southern Africa. Black claims to inequalities and injustice were readily accepted at face value, despite the fact that the *Apartheid* structure had been radically liberalized and transformed in the decade preceding the Transition. It appears that UNTAG personnel had mixed success in making friends with locals, either black or white, but all claimed that they had made good friends with UNTAG members from other countries. The Swiss who served mostly in the settler-dominated south were most impressed by the landscape. A railway station reminded one Swiss of a scene out of *High Noon* (Anon 1990b: 40) and references to Western Movies and the "Wild West" were a frequent theme (e.g., Anon 1990b: 49). This was "Picture Book Africa" to them.

Similarly in the black-dominated north, Jamaican Angelinna Griffin, in charge of the Oshikuku District Center in Owambo, epitomized the spirit of most UNTAG officials. Arriving at her posting in the early afternoon, she hoisted the U.N. flag at first upside down. "I want you to share the pride I feel that the UN is here in this little corner of the earth and that I am here to offer the best that I can in its name. There were half a dozen boys helping me put up that flag and cheering as it goes up." Her prime task was voter registration, a challenge that UNTAG took up "ready to brave the unknown." She initially stayed at the local Mission station that was "built like the old West" (Anon: 1990a). Indeed the "Wild" or "Old" West was to be a constant descriptive metaphor employed by Black and White UNTAG members. The gratitude of the returnees was so heartfelt that it made her "big sacrifice for being here so insignificant." Certainly the local people treated the UNTAG operation with more than due respect and several observers noted that they dressed up in their Sunday best just to register to vote.

Personnel from the Caribbean were particularly eloquent about why and what they were doing: they were returning to the land of their ancestors and felt at home "as the people of Africa shared one common ancestry." So strong was this common ancestry that some of them felt that communication was not a problem despite not sharing a common language because the people were so friendly and welcomed them in their own way: "The satisfaction and fulfillment derived from these simple acts (of smiling and greeting) is not easy to describe and will not be forgotten" (*Untag Journal*. N.d.: n.p.). The friendliness of the "sun-baked faces" many felt was the mission's key to success. Time and again UNTAG personnel, at least those serving in the dangerous north, were impressed by the friendliness of the local people as epitomized by smiling, waving, and greetings.

Others were not so sure. Thus Peter Burke, an Irish policeman stationed in the North, claimed:

It took a while for the people to have confidence in us, but the real turning point came when we got our own UN Casspirs. The Casspir was the most intimidating vehicle I've ever seen, not only the size and the colour, but the design of the whole thing.... To see the big white UN Casspir driving around, that really sent the people a different message. Before that we couldn't keep up with the SWAPO teams, but after that, we could follow them anywhere. Then the people began to realize that there was really a change taking place, and that the UN was there for a reason (Anon 1990a).

Reflecting on his experiences, Major Zareen Khan Khatak of Pakistan observed that "During my six months in Owamboland, I was particularly struck by the constant state of festivity prevailing all around. People appeared drunk on the notion of independence. I am sure they know their responsibilities too. Creating a nation is far easier than building one!" (*UNTAG Journal* 1990: 33).

Most UNTAG personnel fondly recalled the good times they had, especially with fellow UNTAG workmates, during the various national celebrations. As an international bonding exercise it was an unqualified success. UNTAG was unique in that it was the first United Nations operation in which female personnel played a large role. Col. Krish Mehra, the deputy commander of the Indian contingent, was concerned about the way U.N. personnel of different sexes greeted each other with "the three kiss syndrome" and the attractions and dangers of "playing." But he concluded that when in Rome one should do as the Romans do and engage in the various nightly escapades and visits to the thriving restaurants and nightclubs in the larger towns. "Fun parties," he noted, were ubiquitous and frequently entailed weekend trips to resorts. He added that no one in the U.N. ever said No (*UNTAG Journal 2:* 1990: 21).

Host reaction to UNTAG

Interviewing Namibians ten years after the event, the first reaction on being asked what was most memorable about UNTAG was "UNTAG BABIES." This in sharp contrast to UNDP officials who deny that many illegitimate babies were born out of UNTAG liaisons and is based on the naïve claim that few mothers came to them seeking assistance (Hearn 2002). But this ignores the culture of local sexual practices and strategies for seeking support. Even nowadays when visiting Windhoek, the capital, people will point out houses which supposedly served as bordellos during the Transition. Given the relatively high salaries UNTAG personnel were paid, many local women and men saw this as an opportunity to obtain money. That there were many amorous relationships is clear from folklore, cartoons, and court records. From senior officials to the lowest soldier engaged in love affairs. There were jokes about middle-class Whites having to unplug and take their phones with them because their maids would call their lovers in Finland or wherever. It is interesting to note that Simmel regarded the love affair as the ultimate example of an adventure. Indeed, the only adventurer he mentions by name is Casanova. Licentiousness is a common characteristic of tourists and soldiers abroad. Mwanaria, a chronicler of the Kenbatt, raved about the nightlife:

> A night at "Club Thriller" in Windhoek's Katutura township, or one of the many night-spots was quite an experience. The music was good, the beer was plenty and the beautiful girls were willing to gyrate their delicate bodies to the rhythm of music. It was not impossible to witness UNTAG personnel immersed in the merry-making to ease up the pressures of the day and just feel good (Mwaneria 1999: 64).[3]

Indeed a close reading of the Kenbatt chronicles and that of journalists (e.g., Lush 1993) suggests that most assaults on Kenbatt personnel were not by reactionary racist whites but aggrieved blacks and that the root of the conflict was sex and drunkenness in shebeens, the black-run unlicensed taverns (Mwaneria 1999: 65, 68, 94).

After initial apprehension, in which urban legends readily circulated among white Namibians such as that UNTAG's bias was shown in the fact that the condoms they issued their personnel were colored red, blue, and green, the SWAPO colors (Lush 1993: 176; Harlech-Jones 1997: 65), most White Namibians greeted UNTAG's presence with humor. Jokes abounded about the "White West" (in reference to the white vehicles they drove). Stealing UNTAG number-plates became a favorite pastime of White youth to the extent that UNTAG were forced to paint the numbers onto their vehicles. Some locals would also paint an "F" in front of the U.N. vehicle logo. Indeed, one of the local movie hits was the South African comedian Leon Schuster's still-popular *Oh Shucks, Here Comes UNTAG* released in 1990 shortly after Independence and just after UNTAG had left. It portrays UNTAG as being a bunch of naïve foreigners being taken for a number of rides by locals, both white and black. It ends with the Afrikaner Hero having impersonated a number of local people including a Herero woman, and making out with the blonde UNTAG medical officer. Of course this was not how UNTAG personnel saw themselves.

While this apprehension was shared by a minority of local Blacks, most Blacks, especially in the former Operational Area of Owambo, welcomed UNTAG, in some cases with considerable passion. In an important sense local people were also having an adventurous time. An adventurer is, apart from one who seeks adventure, someone who is ready to take risks for personal gain (Oxford Shorter Dictionary); "a person who hopes to gain wealth or high social position by dishonest, dangerous or sexually immoral means" (Longmans); and finally "soldier of fortune, one who engages in risky commercial enterprises for profit, one who lives by his wits" (Concise Oxford Dictionary), but clearly here the line between risk and opportunism is problematic. Risk undermines trust relations, and one of the features of Namibian society that has struck a number of observers is precisely its lack of trust. Clearly the transitional situation was one in which the whole populace was thrust into massively increased opportunities for adventure.

Undoubtedly White apprehension was moderated by the economic benefits they derived from the sudden influx of a large number of relatively big spenders. They filled the hotels and restaurants and chased up property prices beyond the wildest expectations of speculators. Hotel accommodations' prices more than doubled within a few weeks. They also created new employment nodes, especially in the Private Security Industry for Diplomats, Development Workers, and other affluent foreigners who sought protection. But it also had a devastating impact as well as people had to cope with a huge influx of returnees. Practically each house in Katutura, the sprawling black township outside Windhoek, had a backyard squatter, and single mothers became increasingly dependent upon part-time domestic work with unscrupulous landlords not being averse to extracting sexual favors. Meanwhile with the high unemployment, crime skyrocketed as well. Not only did the bulging wallets of UNTAG attract youth gangs, but violent and sexual crimes also escalated exacerbated by the fact that the colonial Police were not trained to deal with such forms of criminality. A journalist who lived there reminisced how self-preservation became a priority epitomized by residents increasingly ignoring victims of crime. "Armed criminal gangs, motivated by joblessness, the lack of an effective police force, and a feeling that independence meant you could do what you liked—also roamed the region, stealing, looting and intimidating the population. But the full-scale rebellion feared by so many never materialized" (Lush 1993: 260). Indeed, "the once amicable, all-embracing township was becoming a jungle. Instead of the ubiquitous *braais* (barbeques) they were replaced by clubs which deprived many of the youth of the chance to socialize and mix with people of all backgrounds." *Braais* were not only broken up now by gangs, but the emergence of "Clubs" catering to UNTAG also undermined *braais* economically so that many households and young women found it easier to simply raise income by getting UNTAG "sugar-daddies" (Lush 1994: 212).

On the uniqueness of adventure

So how is one to understand why the South African troops and personnel who were in the same place and of the same broad demographic profile and education as the UNTAG personnel did not generally see their experiences as an adventure while the latter did? In both it is clear that imagination played an important part. In reality, only a small number of SADF troops actually had "contact" with the enemy, but their officers exaggerated the enemy presence to motivate their troops and to keep up their interest. Most accounts and memoirs emphasize that they had lots of time to think and their thoughts were largely concentrated on what they would do once they got back to the "States," typically girls and get-rich schemes (Strachan 2002; Feinstein 1999).

Perhaps part of the answer lies in the element of flirting with danger, to use the Simmelian metaphor. Unlike UNTAG, the South Africans did not blatantly engage in sex. By all unofficial and popular accounts UNTAG was notorious for its sexual liaisons. By contrast, except for the occasional rape, the records seem to suggest that it was rare that South Africans openly engaged in local sexual liaisons (Fowler 1995: 32). A number of conditions would reinforce such behavior. The South Africans came from a state that, with its strict apartheid legal system, had until recently defined interracial sex as a criminal offense. One of the key arguments made for the abolition of the so-called "Immorality Act" was not that it discriminated, but that the stigma attached to whites convicted under the act was such that they committed suicide. Even after "race discrimination" was legally abolished there were social and cultural pressures that mitigated against such sexual liaisons as the South African Defence Force Ethnology Handbook cited so well illustrated.

But there were other factors as well. Adventures achieve much of their value by being unique. There is a curious dialectic at work. Once an adventure has been had, it is never exactly replicable. It is no accident that alpinists have their eyes firmly set to unconquered peaks and explorers gaze lovingly at the blank spots on the map. UNTAG were clearly aware that they were witnessing what they took to be an important and unique event in world history. The South African troopies, on the other hand, were aware that they might have to replicate their experience with the next military camp they attended. This raises perhaps the most significant difference between the SADF and UNTAG. The former were largely conscripts—they had no choice—while all the UNTAG personnel were there voluntarily. Intrinsic to defining an experience as adventure I suggest is the notion of a willful individual decision to allow accidental events to take place. Another corollary, is that in an adventure, as opposed to a misadventure, there has to be an expectation, no matter how minimal, of a positive outcome. This positive outcome is exemplified in a successful homecoming or reincorporation in one's home society. Indeed, I would suggest that some of the principal reasons for going a journey into the relative unknown are rooted in the homecoming, for this gives the journey its ultimate justification for undertaking it. Participation in adventures are ways of demonstrating to others who we are and what we believe in. It is a status placing activity (Rojek 2000: 37). UNTAG saw themselves and were portrayed as such by the mass media as "Agents of Virtue" and there was a distinct tone of melodrama about how they went about their business. For them Peacekeeping was Theater. An important component of contemporary adventure is the prospect of talking about it. Most UNTAGs were exceptional in their home society, after all, they felt they were part of "making History," unlike the conscripted South African soldiers who, at least in their white social circles, were clearly unexceptional. Their experiences lacked "scarcity value." Most important of all, when they came home they were not treated as heroes or as someone who had had an adventure, and this led crucially to their negative definition because adventure tales are preeminently about successful adventures. It is only now, like Vietnam vets, that former South African troopies are starting to discuss their experiences (Andrew 2001; van der Merwe & Wolfswinkel 2003; Fowler 1995, 1996). UNTAGs could brag about their sexual conquests and adventures to their peers while the South African troopies, if they did have sexual adventures, suffered the burden of not being able to brag about them.

Scheibe (1986) suggests that adventure is central to the construction and development of life stories and that life stories are the major supports for human identities that are of course constantly evolving. In Denzin's terms adventures mark "turning points" in one's biography (1989). People require adventure in order for satisfactory life stories to be constructed and maintained.[4] The need to tell others about oneself is perceived in Western society as a mechanism of individuation whereas, paradoxically, as Foucault observed, it is a mechanism of socialization and subordination. Confessions and similar literary productions like adventures, he would argue, do not mystify but rather establish authoritative "regimes of truth." Confessions of the "border literature genre" and adventure tales create the space within which human types (the terrorist, the animal and bestial) can be located and treated, and the rules of proper confessional and adventure discourse systematically preclude alternative classificatory schema (Hepworth & Turner 1982: 89). But adventures and tales thereof are also searched for what they reveal of "character." Indeed the raison d'etre for many youth organizations having an adventurous slant is precisely to develop "character." "Character" in the bourgeois world, as Goffman (1967: 214–225) would put it, includes a complex of observations and characteristics (in the proper sense of the word!) such as the ability to stand correct and steady in the face of sudden pressures. Some of the major forms of character allegedly revealed in and reinforced by adventures would include: courage, or the quality of envisaging immediate danger yet proceeding; gameness, as demonstrated by sticking to a course of action regardless of setbacks; integrity, or the ability to resist temptation; gallantry, the capacity to maintain proper etiquette; and perhaps most importantly, composure (or poise), sometimes glossed as dignity, or success at sustaining self-control, self-possession, and ultimately the ability to sustain one's bodily decorum, which is then taken as exemplars of "presence of mind."

There is another factor that played a role here. Hans Magnus Enzensburger (1997: 127) has pointed out that nowadays the global elite/bourgeoisie sees it as a human right to distance themselves as far as possible from their own "civilization." The logistics structure of UNTAG clearly facilitated this, and as a result such a condition was fulfilled in Namibia. It allowed participants enrapture about the wilderness while being comfortable, imagining it both as clouded in inaccessibility *and* impressively accessible through superior infrastructure.

But how important was this glow of adventure for the success of the UNTAG operation? David Lush, a young British journalist, offers a unique perspective on the Transition. Unlike most of the UNTAG personnel and the scores of visiting journalists, he lived in the sprawling township of Katatura and associated with the local blacks. The picture he paints from this vantage point is significantly different from the comfortable UNTAG self-image that was sustained by the visiting tarmac-bound journalists. UNTAG personnel settled "into their new surroundings in a manner not too dissimilar to a horde of holiday makers" (Lush 1994: 138). For all their talk of patrolling, UNTAG's presence was largely in the cities:

> In the bars, the UNTAGs continued to fall off their stools and brawl with the locals. In fact, everywhere you turned there seemed to be someone in uniform wearing a blue beret. (Yet) Throughout the 2000 km I traveled during my trip to the war zone, I saw just three UNTAGs; one at Oshivelo (a control post) and two in a Kombi turning into Ondangwa airbase. So much for the peacemakers..." (Lush 1993: 154).

Lush presents a devastating picture of how the notorious para-police, *Koevoet*, would simply ignore the UNTAG monitors and openly display their contempt for them, and how the Police would, when needed, ignore the UNTAG monitors and wade into strikers, while tolerating riots by the pro–South African political groups (Lush 1994: 229). Even after the deployment of UNTAG in Owambo, the South African forces, and *Koevoet* in particular, could still hunt down (alleged) SWAPO guerrillas in what one diplomat described as a "licensed turkey shoot" wiping out several hundred valuable SWAPO cadres. Indeed, Lush provides powerful evidence that when faced with a riot or a massive conflict UNTAG was totally ineffective. Their role, he seems to suggest, was largely ceremonial. The only power they seemed to have was when the United Nations Special Representative threatened to suspend the Transitional Process. Indeed they could not have known or related to most of the populace because most Namibians and certainly those in the "War Zone" hardly spoke English (the lingua franca of UNTAG personnel). For communications most UNTAG personnel had to depend upon interpreters, who were generally in short supply, and had in some cases formerly worked for the SADF. Nor could they appreciate the nature of "Terror," especially its spectral element or the subtle acts of intimidation that occurred in everyday interactions. UNTAG personnel could not know, for example, that certain known policemen were present at public gatherings dressed in ordinary clothes, or that individuals might have been ambiguously threatened with abduction of their children or their crops and houses destroyed by "mistake," or that all the weapons surrendered on demobilization had in fact been accounted for. In short, they could not get to the underneath of life in a seeming tranquil sun-drenched scene.

In retrospect there seem to have been two structural factors that contributed far more to the success of UNTAG, which the literature on the Transition have largely ignored. The first is that the coffers of the South African State were being drained and international bankers were calling in their loans. Economically, South Africa could not afford the war anymore and really had no alternative but to bow out gracefully. Second, I would suggest that far more important in ensuring that the Transition worked was not UNTAG per se but the myriads of international (mostly) white observers who descended upon the Territory to observe and offer advice. These were the people who pushed beyond the tarmac to the outermost nooks of Namibia and threatened to unleash the politics of embarrassment upon the South Africans. Unfortunately no exact figures exist concerning their numbers, but most observers agree that their numbers were substantial. Certainly judging from oral evidence and some written accounts (Carton 2000) they were sometimes in danger but always accorded high status by local people, a situation largely attributable to what Goudge (2003) calls the "whiteness of power." Race played an important if unacknowledged role in both the discourses and practices of power during this transition. Race, especially in Apartheid-ridden Southern Africa, was part of the still-pervasive discourse of superiority and inferiority.

To be sure UNTAG were lulled into a sense of adventure by a number of factors, not least the friendliness of the natives. Having talked it up as an Adventure, they could do nothing but treat it as such. Adventures must be recounted after the fact if they are to be treated as adventures, but unlike adventures of yore, our UNTAG adventures proclaim only what everybody already knows. The visual world of advertising and propaganda both for and against Apartheid shaped how they framed their accounts. Adventures pretend to provide relief from the world of commodities by escaping from the everyday, but ironically it is the trademark of the adventure that plays the decisive role in calculating its value. The effort of UNTAG lay in confirming the make-believe as the authentic. In the final analysis the key condition for an adventure in Simmel's sense, that notion of the freedom of the "Exclave," is sadly missing. It was never really an Adventure as much as a Tour of Duty because if things got bad they could leave. They had an exit option.

But the narratives generated by UNTAG are important beyond what they tell us about Adventure. The eleven-month UNTAG exercise, which ended as planned, and on schedule with Independence on 19 March 1990, only cost some $383.5 million, far less than the original request of $700 million (United Nations 1990: 445). It also enjoyed widespread support, with more than fifty countries providing personnel. Despite having an authorized upper limit of 7,500 military, at its peak it consisted of 4,493 at all ranks, 1,500 civilian police, and just under 2,000 international and local staff. Specifically for the elections, the mission was strengthened by some 1,000 additional international personnel. During its operations UNTAG suffered only nineteen fatalities, mostly the result of road accidents. By these counts alone it is easy to see why it was heralded as a success by senior U.N. officials and this set the tone for later assessments of the Operation.

Conclusion

The task of UNTAG was to assist in transforming a zone of fear and insecurity into a culture of tolerance. Of course creating a civil society takes much longer than a year. Constitution writing and supervising the elections were the easy part. Scholarly assessments of UNTAG read like many "development" documents, so well critiqued by Jim Ferguson (1992) and Emery Roe (1995): narratives based on unquestioned assumptions that develop their own momentum and ignore context and reality. Thus the World Bank study on DRPs that compared Namibia, Uganda, and Ethiopia pays surprisingly little attention to "External Assistance" and then only to discuss strategies and timelines, purpose of support and coordination (Colletta et al., 1996: 60). Like most UNTAG personnel, the World Bank authors claimed to have found the local people abysmally ignorant, supposedly because they had suffered from a long period of social and cultural isolation.[5] There was little "real" information in the Territory and thus information had to be regenerated. They also ignore the role of NGOs and the role of the world economic system, perhaps precisely because they were so concerned to portray UNTAG as an Adventure of heroic proportion.[6]

Christopher Hope, in his satirical novel *Darkest England,* captures the U.N. and South African dynamics and outcomes well, and far better than I could, when he has his narrator muse:

> Africa is not so much a place as an *Adventure,* which they made up as they went along. And truth was never allowed to interfere with the pleasures of certainty. This wise policy ensured their progress in Africa for generations, their ideas so firmly fixed that no subsequent experience was allowed to disturb them. How else does one explain how they succeeded so well and saw so little? (Hope 1996: 162).

Notes

1. This area, which contained over half of Namibia's population, had never been an area of white settlement for a variety of demographic and geographic reasons.
   2. Again our *Frontline* troopie:

> The guy who is lost is the guy in the middle. He can't win. SWAPO comes and says who do you believe, us or South Africa? So he says you, but SWAPO doesn't believe him so they hit him. Then we come and we say who do you believe, us or SWAPO, so he says you, but we don't believe him so we hit him.... We had two cases where the guy says he believes SWAPO. So we wipe him out. He's a terrorist sympathiser (Anon. 1985: 54).

3. He also reported that even before departing from Kenya many young men on their last home leave, hurriedly "took wives" supposedly on the grounds that nobody knew what to expect in Namibia (Mwanaria 1999: 45).
   4. Of course such adventures need not be direct but can be vicarious and have a strong fantasy element.
   5. These figures, Henning Melber points out, illustrate the double standards so characteristic of this era. No one bothered to base any of these guestimates on sound empirical assessments. As it turned out, Namibia's literacy rates were very much higher, much to the embarrassment of the current regime, which tried to market the country to aid agencies as a less developed country.
   6. Indeed after UNTAG most U.N. operations appear to have been much more problematic due to a variety of factors including manipulation by more powerful countries by threatened withholding of funds and a chronic lack of resources (Polman 2003).

CHAPTER 14

A Head for Adventure

Steven Rubenstein

People typically understand adventures as endeavors that are exciting precisely because they are so risky. This experience is so uncommon that people often must escape, or at least take a break from, their ordinary lives in order to achieve it. Characteristically, though, Georg Simmel reminds us that adventures are successful because they intensify, rather than renounce, the tensions of everyday life. "In the adventure," he observed, "the interweaving of activity and passivity which characterizes our life tightens these elements into a coexistence of conquest, which owes everything only to its own strength and presence of mind, and complete self-abandonment to the powers and accidents of the world, which can delight us, but in the same breath can also destroy us" (1971: 193). Anyone who has been in love or who has conducted ethnographic fieldwork knows these feelings. These experiences are intense not so much because they involve risks, but because of the way they can both require—and lead to—moments that simultaneously involve both conquest and surrender. But Simmel is not merely reflecting on a particular kind of experience. His adventurer is one of a number of social types he analyzed in order to show how people who travel—geographically and emotionally—outside the bounds of society are nevertheless driven by and illuminate social forces. The lovers who surrender themselves to their feelings; the anthropologist who leaves her home to immerse herself in the life of another society—all must recognize, sooner or later, that they are acting out a drama not entirely of their own making.

The contributors to this volume are not the first to see the importance of Simmel's project for anthropology (see Murphy 1971; Shack and Skinner 1979). Nevertheless, while Simmel's search for transcendent universals assumed a humanistic conception of culture characteristic of European modernity, it is at odds with the relativistic conception of culture that emerged among anthropologists after the Great War. Thus, while Simmel is supremely attentive to the tensions between the individual and the world, he has little to say about the specific ways these tensions are socially constructed or narrated. But what constitutes an adventure in one society may not in another.

In the context of Simmel's work as a whole, "the adventurer" seems to be a reaction to, and expression of, modernity. Like most European social theorists and philosophers between Waterloo and the Somme, Simmel was concerned that along with technological progress and economic growth, modernity seemed to breed personal and social dislocation and alienation. The adventurer is one example of a person who, through spatial and social dislocation, may actually achieve a transcendence the loss of which modernists typically mourn.[1]

This essay argues three basic points. First, "adventure"—as conceived by Simmel—is not unique or even characteristic of modern Western culture. One counterexample is the Shuar Indians of the Ecuadorian Amazon. Prior to the formation of the Shuar Federation in 1964, most Shuar occupied the tropical montane between the Andes and the Amazon rain forest and relied on swidden horticulture and hunting for food. The loosest of kinship ties linked households, and Shuar had none of the institutions that characterize tribal society, such as sodalities or lineages.[2] Although one could easily apply Western sociological concepts like "mechanical solidarity" or gemeinschaft to Shuar society, it would be a mistake to think of it as characterized by homogeneity and conformity. Indeed, Simmel might have characterized it as a society of adventurers: living in autonomous homesteads scattered throughout the forest and semi-nomadic, Shuar were highly mobile both geographically and sociologically. It would be an even greater mistake to assume that Shuar lived unalienated lives. Although Shuar, like other egalitarian peoples, controlled their material means of production, their notions of personhood and social reproduction revolved around utterly alienable spiritual forces. The centrality of detachment and alienation in Shuar culture is evident in the Jivaroan practice of producing *tsantsas,* the shrunken heads of others (most often, of their neighbors to the east, the Achuar) slain in a raid.[3]

Second, the form that adventures take, and the play of conquest and abandonment identified by Simmel, vary from culture to culture. There are numerous books written by Westerners detailing their adventures in the Amazon. Many of these books hinge on a quest to encounter head-shrinkers and acquire shrunken heads.[4] But did Shuar themselves consider head-hunting an adventure? Thanks to Janet Hendricks, who conducted research within the Shuar Federation in the early 1980s, we have the detailed account of Shuar feuding and warfare by Tukúp, the most famous, and at the time probably the oldest, warrior.[5]

In this essay I contrast accounts of Western adventurers who sought out shrunken heads with Tukúp's account. I argue that Tukúp narrated head-hunting not as an adventure but as a hunt. Among the Shuar the societal forms that correspond most closely to Simmel's conception of adventure in terms of both surrender and conquest are those involving vision quests which, unlike Western adventures, resist narration. Indeed, in contrast to Shuar adventures, what is striking about Western adventures is the way they invite excessive narration.

My third point, then, is to call attention to the way that difference in the amounts of narration signal different dynamics between what Simmel referred to as conquest and self-abandonment. By narrating their stories of head-hunting while representing vision quests in ways that resist narration, Shuar subordinate conquest to self-abandonment. Westerners, conversely, narrated their stories in ways that subordinate self-abandonment to conquest. A discussion of the place of adventure in Shuar society will provide a critical perspective on Western culture Simmel lacked, and thus bring to light certain features of Western adventure he ignored.

Shuar Head-hunting

Tukúp's account begins in his childhood, when his father Kumpánam is killed. Tukúp grows up watching his father's brother-in-law, Pakunt, kill a series of men, because these men made his sister a widow, and caused his nephews to suffer. This story not only establishes Pakunt as a role model; it excuses Tukúp's inaction, for he is just a boy. Not until he joins his brothers on a raid does he become a man—although his older brother Uwek actually kills the victim, Tukúp participates in the raid and shoots the corpse. Tukúp also joins Pakunt and others when they go to kill another man,

Jempe. Pakunt abducts Jempe's widow, and attempts to create an alliance with her brother, Wisúm. But when Wisúm invites Pakunt to visit, Jempe's son kills Pakunt.

When Tukúp learns that Pakunt is dead, he believes the killers will also kill Pakunt's family. Tukúp thus constructs a palisade around his house, and kills the first man who approaches. A second group comes and kills Tukúps's older brother, Uwek. Tukúp and his younger brother Piruchkin take a hallucinogenic infusion, *maikiua,* and decide to avenge Uwek's death. Outside the man's house, the two argue over tactics: Tukúp wants to wait until dawn, but the younger brother is impatient and approaches the house. Just then someone leaves the house to urinate, sees Piruchkin in the moonlight, and calls an alarm. The two brothers retreat.

Tukúp provided Hendricks with accounts of many more raids, both successful and unsuccessful. These stories involve geographic movement, abandonment to forces outside the warrior's control, and attempts to dominate and vanquish another. According to Hendricks, Tukúp is a masterful storyteller, using intonation, repetition, and shifts from first to second person to dramatic effect.

Nevertheless, I suggest that Tukúp's narrative is not one of adventure, in Simmel's terms. We have none of the dreamlike sense of drifting out of one's own time and place that Simmel associates with adventure. And while each specific raid has a relatively clear beginning or ending, the narrative moves quickly from one raid to another, as if each forms part of a larger story. Far from taking us out of the ordinary, they present the life of an ordinary Shuar warrior. Dramatic conflict comes not so much from one man killing another, as from Tukúp's attempts to recruit older or younger warriors as allies, and debates among warriors over tactics. Although the outcome of a raid is uncertain, it is seldom because of purely chance events (such as the moonlight shining on a urinating man). Tukúp instead emphasizes the importance of tactics, planning, and patience. Most raids do not end with the shrinking of the victim's head. Tukúp describes those raids that do result in the making of a *tsantsa* in the same voice and style as any other raid—the only indication that a *tsantsa* was taken is the use of hunting metaphors (e.g., "after cutting it I saw they had cooked it," p. 138). Indeed, based on these accounts it is hard to imagine what difference there may be between head-hunting and any other hunting.

That Tukúp exemplifies the hunter more than the adventurer is not surprising; Philippe Descola remarked that predation is the dominant value of people like the Shuar (Descola 1996: 90). Human predation of course is not merely an end in itself; nor is it solely a means of physical sustenance. It is a social activity that assumes and reinforces various beliefs about relations between predator and prey. As Carlos Fausto recently noted, "Amazonian societies are primarily oriented toward the production of persons, not material goods" (Fausto 2000: 934; see also Santos Granero 1986 and Turner 1995). Fausto thus characterizes Amazonian warfare as a form of "familiarizing predation," in which the production of new persons depends on the appropriation of the capacities and incorporeal constituents of victims (Fausto 2000: 937).

Shuar Vision Quests

The centrality of hunting, both materially and discursively, in Amazonian societies does not preclude the possibility of adventure, but it does mean that Shuar adventures occur within a different frame of reference than Western adventures. Fausto contrasts predation with reciprocity, namely, a system based on circulation and exchange. Although Shuar warfare may be at odds with any principle of reciprocity (see Descola 1996: 90), it is subordinate to a larger system of circulation. This system is composed of *wakán* and *arútam*. *Wakán* has been glossed as "spirit" or "soul," and *arútam* as "vision" or "power," but it is more useful to understand these as forces that take form and direction through interaction with humans at specific moments. The circulation and transformations of *wakán* and *arútam* constitute a metaphysical order that links production, consumption, life, and death. It is within this order that Shuar adventures occur.

Wakán is self-awareness, and comes into being with the birth of new people. In order to become powerful, however, a person must acquire an *arútam* through a vision quest.[6] In the past, boys of about eight years would be taken by their fathers on a three to five day journey to a nearby waterfall, during which time the boy drank only tobacco water. At some point the child would be given *maikua* (*Datura arborea*, Solanaceae) in the hope that he would see momentary visions, or *arútam* (which may take a variety of forms, including animals, a large human head, or a ball of fire). If the boy were brave enough, he would approach and touch the vision, which would then explode and disappear. The boy did not talk about this with anyone, instead he went to sleep. The *arútam* would return in the form of an old man, and reveal himself to be the boy's ancestor, one who lived a long time and killed many people. This old man was the *wakán* that earlier produced the boy's visions. When the old man disappeared, the *arútam* of this *wakán* entered the boy's body and augmented his *kakárma,* or power (Harner 1984: 138–39). As long as the *arútam wakaní* remained in the young man's body, he would express confidence and intelligence, and be able to resist disease, sorcery, and physical violence; if he acquired additional powers, he was invincible.[7] Arútam are alienable. As long as a man possessed arútam, he could not die. It is inevitable, however, that at some point a man's arútam would leave him. If a young man talked about his visions with others, he lost his arútam.[8]

Since warriors on the eve of a raid customarily took turns describing their *arútam* to one another, a young warrior would have to seek new *arútam* in the future (Harner 1984: 140).[9] If he acquired a new *arútam wakán* before the previous one had ebbed away, the remaining power was "locked-in" (Harner 1984: 141). Many adult Shuar men thus repeated their quest for an *arútam* throughout their lifetimes whenever they felt weak, especially since they believed that a Shuar would die in battle if he lost his *arútam wakán.* The accumulation of power, however, required not only the repeated drinking of *maikua,* it required repeated participation in raids. For after inhabiting the same body for three or four years, an *arútam wakán* would begin wandering at night, as the warrior slept. During such nocturnal walks someone else could acquire the *arútam wakán* (Harner 1984: 141).[10]

Beliefs about *arútam* transform the death of a warrior into a productive event. At his death a warrior gives birth to the same number of *arútam wakán* that he had possessed over the course of his life (Harner 1984: 143).[11] Moreover, a great warrior could try to will his *arútam* to his sons, by instructing them to leave his corpse seated on a stool and leaning against the centerpost of the men's end of his house. The sons would then take turns each night to enter the house in the hopes of encountering an *arútam wakán* (see Harner 1984: 168–69).

Arútam are linked with productivity and life; *wakán* are linked with consumption and death. If a warrior died from some natural or accidental cause, his *wakán* became an *iwianch* (Karsten 1935: 373; Harner 1984: 144). According to Harner, *iwianch* start out as analogues of the deceased who relive their former life-history. At that point they are transformed into "true" *iwianch* who continue to exist in a state of hunger and loneliness, though perhaps in a more generalized form. It is still common for adult

Shuar to frighten their children by threatening that this *iwianch* will steal them away. According to Harner, however, Shuar believe that the *iwianch* only wants to play with children, and would not harm them. The *iwianch* represents the opposite of a fulfilling life; it is an ugly spirit consumed by perpetual hunger and loneliness. Indeed, Shuar fear the kind of existence that *iwianchi* represent, more than the *iwianch* themselves.[12] Other *iwianch* haunt the forests and trails, especially at night. These *iwianch* take the form of animals or Spanish soldiers.

If a warrior is murdered his *wakán* is transformed into a *muisak*, a destructive force that seeks to kill the murderer or members of his family (Harner 1984: 143–44; see also Karsten 206, 308, 368). After shrinking a head (through the process Tukúp called "cooking"), the warriors sewed up its lips and rubbed its skin with balsa-wood charcoal to contain the *muisak* (according to Rafael Karsten, this was the principal purpose of head-hunting). Warriors then tamed the *muisak*, and channeled its violent power productively, through various rituals. Through these rituals Shuar articulate the metaphysical world of adventure with the physical world of production and consumption.

The actual transformation of the *muisak* into a productive power occurred through a series of feasts held at the household of the warrior or one of his allies.[13] Repeated performances, rather than the production of new narratives, dominated these feasts. During the first feast, *numpuimartinyu* ("paint with blood"), the slayer (holding the *tsantsa*) and women of the household danced between two rows of men and into the house in order to "paralyze" the threat of the *muisak* (Karsten 1935: 301).

The second feast, *suamartinyu* (paint with Genipa americana L., Rubiaceae, the juice of which stains skin black), took place a few months later. The object of these rituals was no longer to contain the threat of the power of the *muisak*, but to augment this power and harness it for domestic production. During this feast, therefore, the slayer continuously has intercourse with the spirit of the enemy he has killed, whom he meets and converses with, especially in his dreams. Through the ceremonies of conjuration performed, the *wakani* [i.e., the *muisak*] is now turned into his obedient slave, and is obliged to put his superhuman power and knowledge at his disposal (Karsten 1935: 315). This power expressed itself not only directly, in the growth of the slayer's domestic animals and gardens, but indirectly, in the ability of the slayer to instruct with "extraordinary insight" his wives and daughters in their domestic chores.

The third feast, the *einsupani* ("people feast"), with well over a hundred guests, lasted several days. It represented both the final transformation of the *muisak* from a destructive to a productive force and the expression of the fruits of this process. Karsten reported that huge quantities of manioc-beer and meat were served at this feast (1935: 358)—a feat few Shuar are capable of today. On the second day the warriors reenacted the raid and taking of the head, and then drank *natem* (Banisteriopsis caapi, a hallucinogen more commonly used by shamans), hoping for visions of a long life and prosperous future (Karsten 1935: 346).

Violence seems to be at the heart of these beliefs and practices. The murder of one person calls for the murder of others. The decapitation of a head is believed to unleash a fury, but precipitates a series of feasts that legitimate a man's domination of the labor of his wives and unmarried daughters. In a society where biological and social reproduction depend materially on the labor of women (Harner 1975; see also Lorraine 2000, Seymour-Smith 1991), the most productive thing a man can do is kill or die.

There is much drama in this way of life, as Shuar alternate between surrendering themselves to the souls of their ancestors, and conquering the souls of their enemies. Yet in Tukúp's accounts, there is little sense of adventure. The Shuar do not lack adventures, but they are not found in Shuar narratives of warfare. Tukúp's own accounts, however, point to adventures that cannot be articulated: his visions. As he notes, a warrior did not embark on a raid without having had a vision (Hendricks 1992: 149, 165). The acquisition of an *arutám,* and the domestication of the *muisak* at the various feasts, depended on visions.

As Simmel observed, for some life itself may be an adventure (1971: 192). This is, or at least was, precisely the case for the Shuar Indians of the Upper Amazon. But the adventure of head-hunting did not take its meaning from the personal histories of warriors. Cycles of killings extend before and after the life of an individual. As Tukúp's story suggests, a Shuar warrior was born into a world of simmering feuds. Many of his raids avenged killings that occurred before he was born. Similarly, he died leaving a legacy of killings to incite new acts of vengeance. The acts of warriors took meaning from a larger metaphysical order, in which the empty, consumptive existence of the *iwianch* was counterposed to the violent but, under the right conditions, productive forces of the *arútam* or *muisak.* In Shuar culture, the real adventurer is the *wakán,* which may be transformed into any of these three through a warrior's vision quest, killing, *tsantsa* ritual, and death. In these visions, and the world they constitute, the kind of adventure on which Simmel reflects may be found, for it is to these visions that the warrior surrenders himself. His skills in warfare and ritual rely on a metaphysical order that he does not control. Merely to speak of these visions—to assert the power of representation—is to lose power and risk death.

Shrunken heads were important props in this adventure, but not important ends. According to Michael Harner, Shuar men often held on to *tsantsas* after the feast, but as "decorative keepsakes" or as personal adornments, "worn by the head-taker on solitary, meditative walks in the forest...." In this egalitarian society, these warriors did not value their *tsantsas* as indicators of wealth. If a Shuar had not traded away a *tsantsa*, Harner suggests, he could be buried with it (1984: 191). Thus, heads were not to be hoarded or inherited, but maintained as mementos of death. That a valuable object was not hoarded seems strange to Westerners. According to F.W. Up de Graff, an American engineer who joined a Shuar war party in October 1899, "after the feast the heads themselves ... lost their value, as surely as pearls which have died. It is curious that the fanatical jealousy with which they are guarded up to the time of the festival should give place to that complete indifference which allows them to be thrown to the children as playthings and finally lost in river or swamp" (1923: 283).[14]

Western Travelogue

Before the formation of the Shuar Federation in 1964, people, food, *wakáns*, and *arútams* circulated in ways that resisted accumulation. In Western society, however, circulation is often subordinated to accumulation. Adventure narratives written by such men as Fritz Up de Graff, who left New York for Ecuador in 1894, and Leonard Clark, who left California for Peru in 1946, tell of men who traveled to accumulate wealth and knowledge.[15] In writing of their travels among the Shuar and neighboring groups, these men tell of a world beyond their control, a world of rain and snakes, deceitful Hispanics, and dangerous Indians. More importantly, they describe their own odysseys, in which they overcome and conquer these obstacles.

When Western travelogues are compared, what seems to make these stories adventures is a surreal sense of newness. Up de Graff introduces his book by claiming, "since the race of men first began to move over the face of the Earth, the desire for fresh discovery has been strong in the human breast" (Up de Graff 1923: xvii), and early in his book Clark distinguishes himself from other travelers who "weren't real explorers" because they never sought out the unknown (Clark 1953: 37). As Simmel observed, "the adventurer is the extreme example of the ahistorical individual, of the man who lives in the present" (Simmel 1971: 190). Thus, everything is new to the adventurer. Both Up de Graff and Clark discover that electric eels are edible, that Anacondas eat deer, and that Indians are dangerous. Perhaps for this reason Claude Lévi-Strauss found travelogues so monotonous; each one recounts similar experiences in the same way (Lévi-Strauss 1973: 17–18).

Lévi-Strauss recognizes in travelogues something like an Indian vision quest, through which heroic efforts confer great power (1973: 39–41). But he believes that these books and lectures are fundamentally false precisely because of their obsession with newness. Lévi-Strauss's memoir is a meditation on nostalgia, constantly mourning the damage done by European colonialism. He is thus bitter about the ways adventurers delude themselves and their audience into believing that the Old World has not contaminated the New. In other words, he believes that this power comes through misrepresentation.

In contrast to Tukúp's narrative, however, what is striking about Western travelogue is not so much the Westerner's constant sense of discovery, as his compulsion to talk about it. Contrary to Lévi-Strauss's critique, the power of travelogue comes not from misrepresentation but from representation itself. As Erich Auerbach (1953: 6) noted, the need "to represent phenomena in a fully externalized form, visible and palpable in all their parts, and completely fixed in their temporal and spatial relations" is long-established in Western culture (and, as Jacques Derrida (1974) has pointed out, one in which Lévi-Strauss is deeply insinuated). Auerbach starts with Homer's description of Odysseus's scar—a passage of over seventy detailed verses.

This informative but undramatic passage occurs in the middle of a much shorter, more moving, account of Odysseus's arrival home, after decades of war and wandering, when he first makes his presence known to his old housekeeper. This passage is one example of a plethora of passages in which Homer interrupts a dramatic scene to provide a full account of the nature, origin, and history of an object, person, or god (Auerbach 1953: 5). As Auerbach points out, such tangential passages relieve rather than heighten tension (Auerbach 1953: 4), and thus provide the reader with a sense of omnipotence as well as omniscience. The power of this effect, he concludes, is far more important than its accuracy (Auerbach 1953: 13).[16] Similarly, Up de Graff and Clark routinely interrupt their stories with details of the habits of vampire bats, the design of shotguns, or the making of shrunken heads. Whereas for the Shuar to talk of visions is to lose power and invite death, for the Westerner knowledge in the form of speech and writing is power.

This power is never purely discursive—it is intimately connected to material acts. But neither is this discourse merely expressive, describing the material world and events. These authors provide detailed asides to extend western control over hitherto unexploited land. In the introduction to Leonard Clark's *The Rivers Run East* Louis Gallardy (then U.S. Consular Agent in Iquitos) comments,

> In his appendices of the Western Amazon ... the information is 95 percent new. Certainly the wildlife providing hides, feathers and meat, the trees, fishes, medicines and other useful products such as petroleum and gold, will provide priceless material for a study on the possibility of a hundred new exports to the United States and Europe (Gallardy 1953: x).

Yet Clark admits that his detailed attention to fauna, flora, and Indians is itself a ruse, a "cover" (Clark 1953: 4) for the real purpose of his adventure: to find El Dorado, the ancient lost cities of Incan gold. Clark thus proposes to be the hero of an adventure first proposed by the Spanish conquistadores.

If the Shuar subordinate their lives to a larger metaphysical order, Clark subordinates his to a larger metahistorical order. Whereas Shuar saw in their own purposive battles the playing out of spiritual forces over which they have at best tenuous control, Clark believes that his search, which relies largely on rumor and progresses through chance encounters, plays out the grand struggle between savagery and civilization. He is at war with nature itself—"The jungle is your enemy," his guide tells him as they set out, "[i]t will destroy you if it can—and it never sleeps" (Clark 1953: 23). And the great gamble of his adventure is a cold contest against the natives of this jungle: "It was a calculated roll of the dice—my neck against the Indians' secrets...." (Clark 1953: 32).

In fact, Clark's quest for El Dorado was not much of an adventure. The streams and rivers of the upper Amazon were widely known as sites for placer mines. These gold deposits, washed down from veins throughout the Andes, attracted the Spaniards in the first place. Even in 1987, when I first arrived in the area, one could still run into the occasional lucky settler or Shuar who had come to town to sell some gold dust in one of the stores that sold cloth, rubber boots, and fishhooks. In the end, Clark comes across the ruins of some old placer mines, and also discovers Aguaruna Indians operating one. Understanding that these Indians regularly trade gold with non-Indians, he traded what he could, a razor, bullets, clothes (Clark 1953: 323). He returned to the Andes with $16,000.00 worth of gold and a firm belief that he had discovered the source of the Inca's gold (Clark 1953: 338).

In Clark's narrative, the Aguaruna are a means to his extracting gold, and his account of their head-hunting and head-shrinking practices creates both the illusion of a brush with danger and the illusion of security that comes with knowledge. Up de Graff's narrative provides a similar example of adventure motivated by a search for wealth. In this case, however, the Western knowledge of treacherous Indians precipitates a brutal act of violence on the part of the Westerners.

Up de Graff's *Headhunters of the Upper Amazon* starts as an adventure of youth—a young man, fresh out of college, inspired by Henry Morton Stanley's *Through the Dark Continent*, and hoping to find his fortune, leaves New York for Ecuador at the invitation of an Ecuadorian fraternity brother. After a failed business venture involving his friend's family, he returns home via the Amazon, where he seeks to enrich himself first through rubber, and then through gold. It is not just the title of his book, however, that signals his interest in shrunken heads. During his rubber-venture, when he and his companion fail to encounter a single Indian, he stumbles upon a grave. "Here was one of them who at last could not run away from me" (Up de Graff 1923: 93). Intent on taking a "curio," he removes the head and returns to camp (when he leaves his camp to tap rubber, someone—presumably Indians—removes the "curio.")

Much later, Up de Graff joins up with a few other North Americans to pan for gold, when they encounter what they believe are real head-hunters. "We turn to look down-stream.... Sweeping along under the right bank came fifty-five Jívaro canoes in single file, a huge serpent moving with perfect rhythm." The canoes were filled, he discovered, with Aguaruna and Antipas Indians. As the North Americans spoke no Jivaroan languages, and the Jívaro spoke no English or Spanish, they communicated in sign language and a smattering of Quichua.

"We have heard that you have come searching for gold," [the warrior Tuhuimpui] began. "Here on this river are very many Huambizas, evil men, who kill the apaches [Ecuadorians] and steal their women. I myself have talked with the apaches and been to Barranca, and I know. You will be killed, so we have come here to take you back to safety. The Huambizas carry weapons such as yours." (1923: 251)

After offering the warrior a drink, Up de Graff replies that he and his friends are traveling in peace, and can take care of themselves. But then he sees Tuhuimpui conversing with another warrior. Assuming the worst, he consults with his comrades and devises a strategy that confounds any understanding of who his comrades really are:

I have joined my comrades in the shelter…. We decide that we must try and get the whole war-party to go up-stream with us; their help would be invaluable.

In a few minutes I am back talking with Tuhuimpui….

"We are a war-party, come to kill the Huambizas."

Tuhuimpui turns with childlike simplicity to this new story.

"We will help you, then, for they are our common enemy," I answered. "Let us join forces and push on together" (1923: 252–53).

This succession of events is filled with ironies. At first Tuhuimpui identified Up de Graff and his fellows as potential victims of violence, albeit violence at the hands of Huambizas. Up de Graff initially rejected this identification, but then internalized it, fearing that he would indeed be a victim of violence, though now at the hands of Tuhuimpui and his Aguaruna confederates. Finally, Up de Graff identified with Tuhuimpui and described himself and his fellows to the Aguaruna as being precisely what he believed the Shuar to be: "We are a war-party…." On the one hand he made it seem as if his identification with the warriors was cynical and tactical (so, by comparison, the attitude of the warriors was naive and "childlike"). On the other hand, the confusing nature of the text makes the identification total. Who actually said "We are a war-party…?" The following sentences seem to respond to this declaration. Yet each are responses by different people: the first is Tuhuimpui's, the second is Up de Graff's. Thus, the first response seems to be Up de Graff's, and the second response seems to be Tuhuimpui's. What, indeed, is the difference between Tuhuimpui's voice and Up de Graff's? It doesn't matter: Tuhuimpui and Up de Graff were, after all, facing "a common enemy."

Although Up de Graff is convinced that Tuhuimpui plans on killing the North Americans for their trade goods (Up de Graff 1923: 253) and their heads (Up de Graff 1923: 255), the Aguaruna warrior acquits himself quite well. He provides food for his guests ("turkey, pheasants, monkeys, and wild hogs") (Up de Graff 1923: 253) and warns them not to use their shotguns for fear of alerting the Huambiza. Aside from a midday halt, the canoes travel from sunup to sundown. From time to time Up de Graff remarks on how much the Indians must want to kill him, puzzles over why none have attacked, and concludes that they are cowards (Up de Graff 1923: 255). They travel this way for two weeks when the war-party prepares to attack the Huambizas.

On the evening before the raid, the Aguaruna, Antipas, and their five White companions apply war paint. The next day they divide into two groups and attack the Huambiza settlement. The Aguarunas try to sever the head of a woman felled by three spears, but she lives. Up de Graff lends them his machete to quicken the job. Although Up de Graff considers this an act of mercy, he is motivated by his fears—not only that the Indians will kill him, but that they might think him a coward (Up de Graff 1923: 270–74).

After a lengthy explanation of how one shrinks a head, Up de Graff reflects on his party's next decision: whether to separate from the warriors, or continue with them. He provides four reasons to remain with the Indians: first, Tuhuimpui had communicated that they might find gold downstream. Second, the five White men would find it too laborious, paddling upstream without the Indians' help. Third, they feared that if they turned upstream, the Huambizas might retaliate. Finally, "we were anxious to trade the Jívaros out of their trophies—relics of the 'fight' in which we had taken part, gruesome enough to be sure, but nonetheless significant and interesting" (Up de Graff 1923: 285).

Tuhuimpui suggests that each White man captain a canoe, but Up de Graff fears that if separated, they would be easy prey for the Indians. The five men agree instead to divide into three groups (one, two, and two) with Up de Graff, another White companion, and Tuhuimpui in the rear canoe.

> Foreseeing trouble, we had arranged on the night before that in the case of a hostile move on the part of the savages, the signal should be two shots fired in rapid succession. It was agreed that they should not be fired without due cause, but that once the alarm was given each of us was to exterminate the Indians in his vicinity (Up de Graff 1923: 289).

The next day when Up de Graff heard two shots fired ahead and around a bend, he instantly fired into Tuhuimpui's belly as his companion shot another Indian. After dispatching the two remaining Indians on the canoe, they began firing at the canoes ahead. The surviving Indians leapt from their canoes and swam to safety, as shots came from ahead.

Most of the canoes now abandoned, the five Whites met five minutes downstream. One explained why he sounded the alarm: the Indians in his canoe had landed on shore and left the boat. No one had any idea why they had done this, and Up de Graff considered it a "lame story" (Up de Graff 1923: 289) but concluded that since they would have been attacked eventually, it was good they struck when they had the advantage.

Having benefited from the hospitality of Shuar, I sympathize with Tuhuimpui, who—by Up de Graff's own account—always spoke truthfully, and acted helpfully, toward his White guests up to the moment when one of them, with no apparent cause, shot him dead. On the other hand, Tukúp himself observed that attack is a reasonable form of defense (Hendricks 1993: 196). I am more struck by the irony of Up de Graff's justification to his readers. Before describing the raid on the Huambizas, he writes,

> It is at this point in my narrative that I must pause a moment to make a few comments on Jivaro methods of warfare. They are utterly distasteful to the white man—the true white man who is brought up to a code of fair play. The attackers display no bravery, the attacked have no chance to defend themselves. As a cat creeps up behind a bird which is digging up worms, so the Jivaro attacks his enemy. A square hand-to-hand fight he will not entertain. With all his paint and feathers he is, unlike the North American Indian, a coward at heart. I must have given the impression in the foregoing pages that our heart was in the business; but such was not the case.... To have left them to attack the Huambizas alone would have been to stamp ourselves in their eyes as cowards (1923: 270).

Discursively, Up de Graff—like other adventurers—made a rigid distinction between civilized Whites and savage Indians. Practically, however, this distinction dissolves as Up de Graff himself, judging the Indians cowards, fought in a cowardly way.

Up de Graff's cowardly murder of his host does not render his discursive distinction between Indian and White man inoperative. Rather, it suggests that the distinction operates some other way. Both Western and Shuar narratives link power and knowledge, but in different ways. I have already suggested that whereas Westerners narratize their adventures extensively, Shuar do not. Now I want to argue that what is at stake are very different forms of power.

For the Shuar, power is fundamentally alienable and transformable. Not only is it hard to transmit to one's children; it is virtually impossible to hold on to during one's own life. For Shuar, power is connected to forces that easily change and are hard to grasp—*wakán* and *arútam*. The material sign of this power, the shrunken head, is primarily a container for a dangerous and elusive kind of wakán. The shrunken head itself has, within Shuar culture, no exchange or display value.[17] This conception of power was well-suited to Shuar egalitarian culture and classless social structure.

For Clark and Up de Graff, however, power is signified by the accumulation of material goods. Their primary motive is to accumulate wealth in the form of gold, and their accounts of local flora and fauna usually include the possibilities of commercial exploitation. It is not gold, however, that epitomizes Clark and Up de Graff's need to accumulate. Ultimately gold is primarily an exchange value, eventually converted to dollars and spent on other goods, some of which may be collected or invested, others used up. Perhaps all accumulation is haunted by consumption. This fear might be what motivated Lévi-Strauss's critique of travelogue: the more one discovers, the less there is to discover, until discovery itself is all used up. Ultimately, the only secure space for accumulation is in discourse.

We must therefore distinguish between Clark and Up de Graff the adventurers, and Clark and Up de Graff the tellers of adventure. In the tales of their adventures, as Simmel observed, they can be suspended in an eternal present, reliving the moments of discovery over and over. Perhaps this is what compels them to write so much, even at the risk of repeating themselves. One of the most important discursive effects of their work is to transform shrunken heads from signs of anti-accumulation to signs of accumulation by describing them as "trophies," as collectables. These are ultimately lonely men who always need to possess more, and fear losing what they have. They seem to be men condemned to the one form of life Shuar fear most: that of the *iwianch*.

Thus, "adventure" has distinct functions for Westerners and Shuar. Elsewhere Simmel has remarked that the nineteenth century gave rise to a new form of individualism in the West that involved the creation of social and political structures "high above" and with great power over people (1950a: 83). Such products of large-group forces as laws, social organization, and symbols are often antagonistic to and even paralyze individuals (1950b: 96). For such people, adventure offers an opportunity to escape and achieve some sense of power by "conquering" another part of the world.

Shuar society also conjures up super-personal and objective forces in the form of *arútam* and *wakani*. Such powers are indeed alien and alienable, but this does not mean that individual Shuar feel powerless; rather they conceive of power as something that circulates and cannot be insitutionalized. Shuar believe it is the possession of these forces that facilitates human agency; their adventures into the world of spirits are not an escape from the ordinary—they are the basis for action in the mundane world. Unlike Westerners, Shuar who abandon themselves to adventure find themselves empowered, not defenseless.

On the other hand, perhaps a Shuar warrior's sense of conquest is more ambiguous, haunted by a fear of the vengeful *muisak* and vengeful survivors. Their adventures truly take place in a realm they cannot control. If the Westerner's sense of conquest seems more secure, perhaps it is because their adventures typically occur in parts of the world already subject to Western imperialism, if not colonialism (see Stagl 1990).

The Ethnographic Adventure

Ethnographic fieldwork too is an adventure. Like Lévi-Strauss, I'd like to think that my words are truer, more original, than those of men such as Clark and Up de Graff. But even Lévi-Strauss understood the danger of eloquent authority—especially when a Westerner is writing about peoples whose ways of life are threatened by Western colonialism. Thus, Lévi-Struass's structuralist method introduced Hegel's "labor of the negative" to anthropology (see Murphy 1963) and his style of writing in *Tristes Tropiques* invited deconstruction (see Derrida 1973). Elsewhere I too have tried to reflect on the ethnographic adventure in a self-deconstructive way (Rubenstein 2004). This essay ends, however, with an anecdote about one of the ways Shuar narrated my presence when I conducted fieldwork.

Not only powerful Shuar like Tukúp understand the elusiveness and fragility of power and authority; even the smallest demonstrate an ability to question and mock pretensions to power. When I began fieldwork among the Shuar, almost ninety years after Up de Graff's encounters, Shuar were already incorporated into the money economy; they had long ago traded away their shrunken heads and were beginning to measure wealth in terms of the accumulation of land (which could be sold only to other Shuar, and was used primarily for hunting and gardening) and cattle (which are raised for sale on the market). To most Shuar I represented the wealth of the United States; my camera alone cost as much money as the average household might spend in a year, and I had a seemingly continuous supply of money, besides. At the same time, I was needy, depending on them for food and shelter. While Shuar shared with me all that they owned, I often had to explain that I could not share with them what I owned—I had only one camera, for example, and needed to make sure my supply of batteries lasted. During the day I took photographs of Shuar and recorded their words on film and cassettes I would take home with me. In the evening, as I spread out my stock of toiletries and lathered up my face over a small basin, little children would run up to me, scream *"Iwianch! Iwianch!"* and run away amidst peals of laughter.

Notes

1. Given this argument, I agree with Bruno Latour's contention (1993) that post-Enlightenment Europe is not "modern."

2. In the late 1940s Catholic missionaries began organizing nucleated settlements, and leaders of these centros developed political structures to coordinate their activities and to promote their interests. This process culminated in 1964 with the formation of the Shuar Federation (see Rubenstein 2001).

3. Anthropologists have used the term "Jivaroan" to refer to a number of groups speaking closely related languages and characterized (or formerly characterized) by similar subsistence and settlement patterns, and social and political organization, such as the Shuar (Harner 1984; Hendricks 1993; Rubenstein 2002), Achuar (Descola 1994, 1996a; Gippelhauser and Mader 1990; Taylor 1981), Aguaruna (Brown 1985), and Shiwiar (Seymour-Smith 1988). Euro-Ecuadorians use the word "Jívaro" pejoratively, as a synonym for "savage."

When referring to specific groups such as the Shuar I do not use the word "Jivaroan" for two reasons. First, Shuar consider the term an insult. Second, each group has a different experience of colonialism and consequently a different understanding of their own identity. After over a century of colonialism, it makes no sense to collapse these groups into one "academic taxonomic ideal."

Since this essay treats travelogue among the Huambiza and Aguaruna as well as the Shuar and Achuar, I use the word Jivaroan only to refer to all groups.

4. The first scientific paper on *tsantsas* was presented before the Anthropological Society of Paris in 1862; the last documented tsantsa feast was in the late 1920s.

5. Ethnographers of the Shuar typically distinguish between feuding, which is among people of the same group and not for heads, and warfare, which is between people of different groups and for heads. Hendricks argues that Tukúp's identity as Shuar or Achuar is unclear, and that the boundary between feuding and warfare is also unclear (1993: 18). Hendricks provides different estimates of Tukúp's age, sug-

gesting he was born between 1908 to 1930. She estimates that he killed twenty men, although some of Hendricks's sources claimed he killed as many as one hundred (1993: 19).

6. In the past, a girl's vision quest was typically combined with an initiation rite at first menstruation (See Karsten 1935: 236).

7. This discussion of the *arútam* and *arútam wakaní* calls attention only to these beliefs in the context of male discourses on warfare. Mader argues that "arútam" may be glossed as "power," but is in fact a polysemic term whose meaning changes as different Shuar use the term in different contexts (1999: 163–169). Women too may seek and acquire *arútam* by taking *natem* and tobacco juice; although like men women may associate the resulting power with violence, they also associate it with fertility in the garden and in the family (1999: 314 ff.).

8. Shuar believed that it took up to two weeks for the force of the *arútam wakaní* to depart completely.

9. According to Taylor, Achuar believe that the *arútam wakaní* departs immediately upon killing an opponent (1993: 661).

10. According to Clastres (1989: 209–210), the belief that a warrior's power is ephemeral is common in the Amazon.

11. According to Taylor, Achuar believe that the dead do not automatically become *arútam wakaní* but must be instructed by mourners through "soul songs," or anent (1993: 664).

12. According to Harner, after existing in as an *iwianch* for the equivalent of a human lifetime, they transform into giant butterflies or moths called *wampank* (Harner 1984: 150–151). According to Taylor, different parts of the body have their own *wakan,* which upon death transform into animals (Taylor 1993: 662).

13. This summary comes from what is perhaps the most detailed published account of the last known *tsantsa* feasts, which Rafael Karsten observed in the 1920s in Chiguaza, just south of the Pastaza river.

14. The accuracy of Up de Graff's report is by no means certain. My concern here is with what Euro-Americans believed about *tsantsas,* and the significance of such beliefs.

15. Moreover, the circulation of the books is subordinated to accumulation through the marketplace—on the one hand, authors and publishers accumulate money (and authors may accumulate fame), and buyers of books often accumulate libraries.

16. Auerbach contrasts *The Odyssey* with the Hebrew Bible. He intended to show how the same methods of analysis could apply to both secular and religious literature. I suggest that the crucial difference between these works is that one is Western and one is non-Western (although it has been appropriated and reinterpreted by the West).

17. Bennet Ross 1984 and Steel 1999 describe how they came to have an exchange value in contact with European and Euroamerican markets and trade networks.

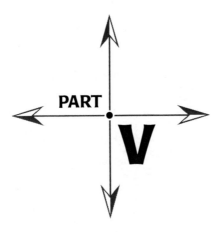

PART

V

Bringing Adventure Home

Riding Herd on the New World Order: Spectacular Adventuring and U.S. Imperialism

Keally McBride

"Adventure has the gesture of the conquerer, the quick seizure of opportunity, regardless of whether the portion we carve out is harmonious or disharmonious with us, with the world, or with the relation between us and the world." (Simmel 1971, 193)

It was the photo opportunity that everyone still remembers. President Bush welcomed the aircraft carrier Abraham Lincoln back in port. He personally flew the plane accompaniment, landed in a jet fighter, and strutted in pilot garb under the banner, "Mission Accomplished." Those who did not support the war, or President Bush in general, were outraged. How could he say "Mission Accomplished" when Iraq was obviously a bloody mess? The liberal media spawned a series of commentaries on the event, *The Village Voice* speculating that the true goal was to present Bush in a uniform that tightly framed his crotch displaying his macho bravado to a nervous public. *Harper's Magazine* reported on the extra expense incurred for Bush's landing to be staged. The pilot wavered about how long the President actually controlled the plane after it was pointed out that he was not licensed to fly the aircraft. Instead of a substantive debate about U.S. military imperialism by those who were opposed to such activities, we had a scuffle about whether the depiction of President Bush was true to life or a staged hoax. We can lift the curtain to reveal the truth behind the image! Mission accomplished indeed.

It is crucial to recognize the fact that the representation of military activity as a spectacular adventure is particularly useful for selling U.S. foreign policy to American citizens. The Bush administration has discovered that spectacularizing the military adventure is an effective way of normalizing what is actually a radical departure from the previous international order and minimizing resistance to its actions. The tropes of adventure also provide a sense of security and relieve the United States of much of the responsibility for its actions in the world. Thus in several ways, the image of military adventuring has become a central aspect of presenting and maintaining U.S. foreign policy.

Today, the exercise of U.S. power is elaborately staged and carefully choreographed and publicized as an unplanned adventure. Considering the current animosity between France and the U.S., it is a delicious irony to consider how George W. Bush and Karl Rove fit into the long tradition of using display as an instrument of power, along the lines of Louis the XIV and Napoleon (Auzepy and Cornette 2003). Opponents of militarism and imperialism can only stand with their mouths agape as President Bush swaggers in uniform, compares other world leaders to beasts, and states that we do not need the permission of others to invade foreign countries. To point out that these actions are macho, inappropriate, and reckless is only to state the obvious rather than provide any substantive critique. How does one go about unmasking what is in plain sight, not only that, but carefully and deliberately calling attention to itself?

If television cameras can follow the ground troops during invasion, if the camera follows Bush as he jumps out of the jet fighter, the message is clear. We have nothing to hide! We are not ashamed of what we are doing! To turn military operation and foreign policy-making into a spectacle, a well advertised and apparently open and honest exercise of American might, does give the whole enterprise a sense of unreality. But even more significantly, I argue this spectacle also obscures that which it illuminates. Michael Rogin's essay, "Make My Day! Spectacle as Amnesia" examines the Reagan and Senior Bush administrations and concludes that that spectacle can accomplish what Poe's antihero in *The Purloined Letter* did: it hides things in plain sight. "Political spectacle in the postmodern empire, in other words, is itself a form of power and not simply window dressing that diverts attention from the secret substance of American foreign policy" (1993: 500). The question is, how does the display of American imperialism serve its goals?

Interestingly, the spectacle only works to achieve cultural unity when it is recognizable and routine. While consumer capitalism has promoted a culture based upon spectacle for more than one hundred years (Benjamin 1999; DeBord 1994), recent fears about security make spectacle an especially potent element in politics. As others have pointed out, the terrorist attacks in September 2001 were designed to be a spectacular strike at the symbols of American power. The nature of terrorism is such that it is calculated to interrupt normal experiences, and to make citizens question the certainties of their everyday lives.

Providing security in this situation demands a reassertion of normalcy and predictability. This is one of the first paradoxical but useful aspects of presenting the spectacle of adventure. Although adventure according to Simmel is supposed to be a break from the routine, today we all know what adventure "looks" like, and it is remarkably easy to signal the liminality of adventuring. Adventure is ideally an exceptional, surprising circumstance; when it becomes spectacularized it becomes relatively predictable and hence easily deployed for strategic effect. Figures cling to crevices in stone cliffs, silent so we can hear the wind swirling up around them from the yawning gorge beneath. Water splashes, snow flies, sweat drops, muscles flex, as Zen-like concentration and precision is displayed in triumph over the obstacles. While the element of chance is clear in actual adventuring, the depiction of adventuring is quite predictable. Hence, visions of adventure contain tropes that make it easy for us to project recognizable roles, expect particular outcomes and behaviors, and perceive a particular form of obstacle or enemy.

The spectacle of imperial adventure is one response to terrorist activities in the United States which can provide U.S. citizens a stronger sense that we had returned to programming as it is normally scheduled. The spectacular, hence predictable and comforting, nature of military adventuring is a crucial element in providing the psychological security sought by democratic citizens, even if it appears that the action of military adventuring has not provided an increase in national security as is conventionally understood.

The spectacle of adventure provides for the translation of U.S. foreign policy into readily recognizable images and archetypes, hiding its truly radical nature. There is no gap between spectacle and content to be unearthed here. We appear to be asserting claims of dominance of a world historical nature. In fact, we are asserting claims of dominance of a world historical nature. What needs to be understood is how the spectacle of adventure allows the U.S. government to pursue imperialist foreign policies and minimize resistance. The display of power can obscure its true nature and distract potential rivals, as court culture at Versailles so clearly demonstrated.

One of Rogin's fascinating observations is that spectacle can serve as a sort of amnesia; we present things prominently in order to promote a form of forgetting. "Since amnesia means motivated forgetting, it implies a cultural impulse both to have the experience and not to retain it in memory. Political amnesia signifies not simply memory loss but a dissociation between sensation and ego that operates to preserve both." (1993: 507) The end of the Cold War presents an unprecedented problem in U.S. foreign policy. How can we act as the world's sole hegemon without appearing to be the bully on the block?

Bush's unique style seems to have struck upon one way of achieving this goal—through spectacular adventuring. The trope of the adventure allows him to exclaim, "We are going to ride herd on the peace process" in response to queries about violence between Israelis and Palestinians, and sneer "Bring 'em on!" when asked about guerilla attacks upon American armed forces in Iraq. He indicates how he truly views the role the United States plays in the world, as the power that guides all others, that engenders jealousy and is yet so strong as to be immune to challenge. At the same time, he declares, with equal sincerity in his 2004 State of the Union address, "America is a nation with a mission—and that mission comes from our most basic beliefs. We have no desire to dominate, no ambitions of empire" (*New York Times,* 21 January 2004). We act as an empire, yet simultaneously deny the role.

The Bush administration has been able to capitalize in particular upon the tropes of adventure as unexpected and liminal. Seeing U.S. foreign policy as a military adventure occludes the relationship between the U.S. and the rest of the world. If the U.S. is the adventurer, the rest of the world is defined as a hostile, unknown environment. Adventurers face obstacles as they present themselves, rather than intentionally seeking danger. Adventurers are willing to suspend the rules in order to survive. Seen in terms of its foreign policy, the U.S. walks its lonely road in a hostile, unknown world. When provoked or attacked it can rise to the occasion. It is capable of meeting the demands of the hostile environment that it has had no hand in making. But what it really most wants is to be left alone to discover its destiny. We do not deny that we are exercising power, only that we planned to or enjoy doing so.

It is no coincidence that Bush invokes the cowboy metaphor in reference to the Middle East peace process, or that he always entertains foreign dignitaries at his ranch in Texas. It is impossible to say whether the Yale-educated Bush feels a true allegiance with cowboy culture. Some statements from his campaign suggest that he did see the world even before his election and terrorist attacks in the stark terms of the adventurer embarking into an unknown world. Frances Fitzgerald reports that during his campaign, when asked about foreign policy and defense, Bush replied, "When I was coming up, with what was a dangerous world, we knew exactly who they were. It was us versus them, and it was clear who the them were. Today we are not so sure who the they are, but we know they're there" (Fitzgerald 2002: 84).

Simmel observed that, "We speak of adventure precisely when the continuity with life is disregarded on principle—or rather when there is not even any need to disregard it, because we know from the beginning that we have to do something alien, untouchable, out of the ordinary" (1971: 189). The stage for the spectacle of American adventuring was set on 11 September 2001. Clearly, normal rules were suspended and it was asserted that the time had come for exceptional bravery and action. What makes this war different from other wars however is that there are no clear enemies and no delineated battle zones. The War on Terror instead is framed as a U.S. military and intelligence adventure that must reach into every crevice of the globe. On 29 September 2001, President Bush proclaimed, "Our war on terror will be much broader than the battlefields and beachheads of the past. The war will be fought wherever terrorists hide, or run, or plan" (Roth 2004: 2). Similarly, Secretary of Defense Rumsfeld characterizes the mission of the United States military in terms that could have easily come out of a description of an adventure film: One goal is "to deny our enemies sanctuary, making sure they know that no corner of the world is remote enough, no mountain high enough, no cave or bunker deep enough, no suv fast enough to protect them from our reach" (Rumsfeld 2002: 26). The reference to SUVs is startling here and perhaps not accidental. The image of adventuring commonly invoked in SUV commercials bleeds into the description even before the make of the car appears in the list of obstacles to be surmounted.

The adventure is presented as a heroic response to terrorist attacks on 11 September 2001. Yet the Bush doctrine of military buildup, preemptive strikes, unilateralism, and world dominance was formed largely before he took office, or even before he was a candidate for office. In 1992, when

Richard Cheney was still the Secretary of Defense in the first Bush administration, he developed a new plan for foreign policy in the post–Cold War era. In March, a draft copy of the Defense Planning Guidance for fiscal years 1994–1999 was leaked to *The Washington Post* and *The New York Times*. The document proposes the primary goal of the United States in the coming decade is:

> to prevent the re-emergence of a new rival, either from the territory of the former Soviet Union or elsewhere, that poses a threat on the order of that posed formerly by the Soviet Union. This is a dominant consideration underlying the new regional defense strategy and requires that we endeavor to prevent any hostile power from dominating a region whose resources would under consolidated control, be sufficient to generate global power (*New York Times,* 8 March 1992).

Additionally, the brief calls for the U.S. "to address sources of regional conflict and instability" and to spread democracy and open economic systems. The way to create and maintain this new world order is to assert undeniable U.S. hegemony, through economic, military, and political capacities. The U.S. must capitalize on its position as the world's sole superpower to make sure that no other players can enter the same weight class. The United States must become a power so great that it is impossible to contest, similar to Hobbes's Leviathan who exhibits strength multiplied so many times over the individual citizen that he is able to create peace from a state of war. The guiding assumption of the Cheney foreign policy is that the only way to achieve stability in a unipolar world is to establish and maintain world dominance.

But increasing military capabilities abroad and interfering in other regions without being asked to do so is difficult to justify to American voters, who tend to like isolationism. Even at the brink of the invasion of Iraq, which was supported by a majority of American citizens, one poll found that only 16 percent of Americans thought the United States "should continue to be the pre-eminent world leader" (Todorov 2003). Furthermore, accelerating defense spending without being spurred on by close competitors is also difficult to justify.

The Cheney plan for world dominance was forgotten, even mocked, during the Clinton administration (Armstrong 2002). Yet policy pundits who were soon to become central to the Bush Jr. White House resurrected the principles of the 1992 plan and presented it in 2000, before the election. In September 2000, The Project for the New American Century issued a manifesto, "Rebuilding America's Defenses: Strategy, Forces and Resources For A New Century." Key administration figures such as Paul Wolfowitz, Lewis Libby, and Stephen Cambone helped put together the report, which became the blueprint for defense spending and foreign policy objectives for the next administration.

In response to the Clinton years, the report pronounces American hegemony threatened. "Without a well-conceived defense policy and an appropriate increase in defense spending, the United States has been letting its ability to take full advantage of the remarkable strategic opportunity at hand slip away" (Project 2000: ii). Arguing for dramatically increased levels of military spending, the report points out that the projected surplus in the U.S. government budget "removes any need to hold defense spending to some preconceived low level" (Project 2000: iii). Bush's campaign platform was to return the surplus to the hands of taxpayers, so this was not to provide the rationale for increased spending. Promised tax cuts helped deliver the presidency, but the strategic defense planning initiative could be implemented under a different logic. Instead, the War on Terrorism provided the stage by which military spending and military action could be pursued. How can an imperialist policy find acceptance with isolationist voters? By making it appear exceptional, unplanned, and unwanted.

In 2002, as required by law, the Bush White administration released "The National Security Strategy of the United States of America." While it is positioned as a response to the War on Terrorism, all the major components of the Strategy were elaborated in the 2000 report by soon-to-be members of the Bush foreign policy team. The plan was already in place, but the exceptional circumstances of the terrorist attacks allowed the Bush White House to begin to pursue them in full sight. Here the supposedly exceptional nature of the adventure allows us to forget that the goal of world dominance was already in play.

The exceptionalism of adventure also signals a suspension in normal modes of existence. In the case of U.S. foreign policy, this suspension is the international order, represented by the United Nations, and the domestic rule of law according to the U.S. Constitution. As Kenneth Roth observes, adopting the rules of war to fight the war on terrorism, a war without any clear boundaries, enables the administration to suspend the normal regulations of law enforcement rules. Instead, the much more loosely defined and erratically enforced international human rights law becomes the operating guidelines (Roth 2004).

Removal from everyday life also signals a lapse in time. Simmel notes, "The adventurer is also the extreme example of the historical individual, of the man who lives in the present" (1971: 190). Staging U.S. foreign policy as spectacular adventure and as a reply to a temporary, immediate threat allows Bush to embrace, even advertise the unprecedented nature of his actions abroad and at home. Adventure suspends the historical continuum and demands action outside of the normal course of events and expectations. Current U.S. foreign policy is seen as necessarily exceptional rather than radically imperialistic. At the same time, the implication is that once the War on Terrorism is won, everything will go back to being the way it was. History will not proceed directly from this era, and therefore there is no need to be concerned about the potential ramifications of these actions. While theorist Giorgio Agamben has pointed out that we have entered "the state of exception" in which the exceptions to the rules have become the rules of the day (2005), the justification for this transformation has largely been grounded in narratives of adventuring.

In addition to the exceptionalism, of action, time, and space in the adventure, a different relationship between adventurer and environment is also established. Knowledge is limited, but intuition is all. Knowledge, rationality, and intelligence will not provide the guideposts for the true adventurer. "In the adventure, we proceed in the directly opposite fashion: it is just on the hovering chance, on fate, on the more-or-less that we risk all, burn our bridges, and step into the mist, as if the road will lead us on, no matter what. This is the typical fatalism of the adventurer. The obscurities of fate are certainly no more transparent to him than to others; but he proceeds as if they were" (Simmel 1971: 194). Rumsfeld explains the unique challenges of American defense in remarkably similar terms: "To defend our nation against the unknown, the uncertain, the unseen, and the unexpected" (2002: 23).

However, this is precisely where the Bush administration has become tripped up in their depiction of the U.S. imperial adventure. Rather than clinging to the script of adventuring, they strayed into the territory of trying to justify decisions and actions according to concrete intelligence. The intelligence about Weapons of Mass Destruction has proved to be incorrect. In the 2004 State of the Union Address, President Bush returns to the intuitive strategy of the adventurer now that intelligence had been discredited. Like the adventurer who moves forward into an unknown future with nothing but faith in his own abilities to sustain him, the United States also moves forward in an uncertain future faced by a possibly hostile world, nonetheless knowing that we will prevail.

> My fellow citizens, we now move forward, with confidence and faith. Our nation is strong and steadfast. The cause we serve is right, because it is the cause of all mankind. The momentum of freedom in our world is unmistakable—and it is not carried forward by our great power alone. We can trust in that greater power who guides the unfolding of the years. And in all that is to come, we can know that His purposes are just and true (Bush 2004).

The imperial adventure of the United States, guided by faith is thus blessed and protected by a higher power. We move not through calculation and intelligence, but rather faith. Here the adventure story clasps hands with election-year politics, invoking another fusion of military adventure and faith—the Crusades.

The other implication of this unique relationship between adventurer and environment is that the adventurer is responding to a world that is unknowable, and also uncontrollable. The adventurer is responsible for responding to his environment rather than shaping it. This implication allows the Bush administration to deny their role in the shaping of the New World Order and position itself in the starring role rather than admitting to being the director, writer, and producer of the entire affair.

The final characteristic of adventure to be investigated here is Simmel's observation that although the adventure itself signals an interruption of the everyday, it also defines the true, essential character of those involved. Adventure is where "the unifying core of existence from which meaning flows" is revealed (Simmel 1971: 191). It is not by creating a spectacle of oneself, but through being tested by the unknown and uncontrollable that a person discovers her true talents, strengths, and character. In this sense, the U.S. imperial adventure establishes the true nature of America at this historical juncture, not through our self-proclamations or posturing, but rather in the response to the demands of the world. Consider Bush's State of the Union Address again:

> For all Americans, the last three years have brought tests we did not ask for, and achievements shared by all. By our actions, we have shown what kind of nation we are. In grief, we found the grace to go on. In challenge we rediscovered the courage and daring of a free people. In victory, we

have shown the noble aims and good heart of America. And having come this far, we sense that we live in a time set apart. (Bush 2004).

The image that emerges from the smoke and haze is a valiant adventurer. A man, yes a man, who breaks the rules in order to defend justice. A man who rescues women from bad men. The image of unveiled and freed women pervades administration statements about the war on terror and its liberating effects. Like the cowboy adventurer, we as a nation will not settle down to marry these newly liberated women, or actually serve as Sheriff in the town now that the villains have been deposed. We will grace some of the competing candidates for restoring order with our official blessings as we ride out of town seeking our next adventure. As one White House official explained after Bush promised to "ride herd on the peace process" in the Middle East, this doesn't mean extended engagement or attention in the details of the region. "The President will ride herd over the two sides, but he's not down in the weeds with the herd" (CBS Morning News, 11 June 2003).

The tropes of adventure allow the Bush administration to present their policies as unplanned, necessary, liminal, thus substantively avoiding criticism. But the means of presenting the spectacle of U.S. empire to the population has its demobilizing effects as well. As an audience member it becomes difficult to tell definitively whether the images are the truth or a lie. We know that taken out of context, some images lie. We know that we are not "seeing the whole" story in the American occupation of Iraq. How are we to confidently assert the difference between image and reality? Upon what grounds are we to evaluate whether the adventure is a successful one or not? Or even more importantly, is it the adventure of a hero or a rogue?

The spectacle of U.S. military adventuring makes it difficult to act politically. In a consumer culture the most frequent exposure any of us have to archetypal adventuring is on screens—both large and small. Zygmunt Bauman (2002: 158) cites a study by Jacques Attali that suggests at current rates of production, sales, and viewing habits, more than two billion TV sets will be switched on at any one time by 2012. Bauman argues this fact has major ramifications for cultural analysis, "With the sun never setting on more than 2 billion switched-on TV screens, the world seen is the world as-seen-on-TV. There is little point in asking whether what you see on TV is the truth or a lie" (Bauman 2002: 160).

Yet the spectacularization of adventure, while it might make the actual experience of adventuring seem "more real," also adds to the dreamlike quality to adventuring in general. Simmel noted that the interruption of everyday context provided a sense of unreality to adventuring, "a remembered adventure tends to take on the quality of a dream" (1971: 188). This unreality is only further perpetuated by the adventure's existence as a spectacle. In particular, those who view adventuring as a spectacle engage in it as a suspension of their everyday day life, and see it as an elaborately staged entertainment for their enjoyment and escape. And very frequently, these adventures depicted in our culture are not real, and are staged for the benefit of viewers at home. Reality television adventure shows, commercials, chase movies, and extreme sports thrive by simulating adventure. However, bombing, invasions, land mines, arrests, and explosions are not orchestrated for the entertainment of the viewing audience. Nevertheless the depiction of wars on television and the conscious posturing on the part of President Bush to fashion himself and the war on terror according to the spectacles of adventure in popular culture enjoy this same dynamic amongst viewers. Domestically, when images of battle float onto our television screens, it is difficult not to conflate those images with ones from *The Terminator* movies or the television series *24*. It is easy to experience, even if not consciously, the images as unreal adventuring staged for our viewing pleasure, rather than as very real events that are changing and ending the lives of people around the globe.

I have already suggested how spectacle reinforces the particular terms of the Bush foreign policy. Those who watch the spectacle of U.S. imperialism then join together in a collective understanding of this episode in U.S. history as necessary and contingent. Just as individual viewers end up being unsure whether or not to trust their own eyes, and thus their individual orientation and ability to act, the spectacle as a cultural form has other political implications. Spectacle provides a way that people can be unified, yet still essentially alone. "Spectators are linked only by a one-way relationship to the very center that maintains their isolation from one another. The spectacle thus unites what is separate, but it unites it only *in its separateness*" (DeBord 1994: 22). This insight reveals the unique political utility of spectacle. We can all metaphorically rally round the troops, be informed, listen to the same victory speeches, and yet not do so collectively. The diffusion of images and information is universal, but the media that accomplish the transmission are profoundly isolationist. When adventure becomes a spectacle it psychically unites people, but also serves their social and physical isolation. Hence, the spectacularization of adventure is a way to provide security of a collectively perceived script, without engendering collective action. The problem is that even those who do not believe the script that is being presented find themselves quarrelling over its production, rather than the central narrative. It is difficult to imagine how to displace the adventure story that is so well known and loved with something more uncertain and less triumphant.

Conclusion

What is curious is how resilient the world is becoming at resisting this image. Does this signal the end of American hegemony through cultural forms? The image has been deployed and its ability to translate U.S. imperialism into a mode of entertainment and an interruption from the banality of everyday life has been convincing indeed. Yet the rest of the world doesn't seem to be buying into the central narrative. Consider the refusal of even Mozambique to vote to approve the U.S. invasion of Iraq in the Security Council. Even given the U.S. dominance through political, military, and economic means, administration officials are having an extremely difficult time extending the narrative of adventure to include other players.

U.S. military might, well, this is unquestionable no matter whether the projector is rolling or still. So why is it that the most fearsome military and largest economy in the world is having problems finding allies? Certainly, there is an aura of bad faith in our spectacle, that perhaps even we are not convinced by the story, so why should anyone else believe it? This might be due to the nature of the script, in addition to the dubious ending that awaits us when the film has been turned off and lights come on. The adventurer in Bush's foreign policy is a lonely one, a unilateralist rebel who doesn't need friends or allies to survive, or even win. As DeBord (1994) noted, spectacle has a way of dividing its audience, as well as making them inactive. If watching action is all that the script requires of most of the world, then generating action amongst the viewers is bound to be more difficult. Even a spectacle that brilliantly hides facts in plain sight might have its downside.

That the Bush administration is placing a great deal of faith in their images is undeniable. Electoral and legislative politics have also become subservient to militarized spectacle. Consider House leader Tom DeLay taunting Democratic leadership by crowing: "To gauge just how out of touch the Democrat leadership is on the war on terror, just close your eyes and try to imagine Ted Kennedy landing that Navy jet on the deck of that aircraft carrier." (*New York Times,* 29 July 2003). Similarly, a Democratic voter in Iowa, when asked about which candidate he would support to run against President Bush, decided that image was the decisive factor. It needed to be "someone who will look impressive enough at the helm of an M-1 tank" (*New York Times,* 4 August 2003).

Opponents to the U.S. military are also getting more effective at fighting this imperial power on its own terms. The prominence of videotapes in taking hostages, stating demands, and presenting killing is not accidental. Their videos have a grainy, unmediated quality that implicitly questions the "realism" of our own media. In April 2004, photos of flag-draped coffins were published to the fury of Administration and Pentagon officials, showing some of the bodies of the soldiers on their return to the United States. The body count is not part of spectacular adventuring; the fact that such images are being so closely regulated demonstrates how self-consciously the spectacle is being crafted.

CHAPTER 16

Adventure and Regulation in Contemporary Anthropological Fieldwork

David Stoll

To establish ourselves as professionals, anthropologists have long tended to downplay the adversities we encounter—distrust, opposition, calamity, irreducible ambiguity—in order to protect the credibility of our research. Even the rather attractive category of adventure usually finds a place only in our memoirs or popular treatments, not in peer-reviewed articles and books. Candid portrayals of adventures and the complications they leave behind could undermine the air of impartial authority for which most of us strive. Now that cultural anthropology has become absorbed in how our position as observers affects the knowledge we produce, the anthropologist as adventurer becomes pertinent in a new way. But anthropologists who boast of adventures in some contexts will have good reason to deny them in others.

The reason is a basic ethical dilemma in anthropological research—the power differential between ourselves and many of our subjects and what this can mean in a crisis. Perils that we can recount as an adventure, because we were able to escape, can spell death and destruction for our subjects, because they could not. When anthropological decision-making imperils the people we study, we violate the ethical imperative of doing no harm. Heightened consciousness of this and other inequities has led to a new era of regulation in anthropology, in which there is even less room for adventure than before.

One expression of the current regulatory era is the human subjects protocol, an externally imposed form of regulation to protect the subjects of research from the researchers. However many reasons anthropologists may find to study adventure, having an adventure ourselves entails risks that violate human subjects requirements. A second, more internalized expression of the new regulatory era is epistemological shame, a systematic distrust of Western claims to knowledge that is often accompanied by reverential attitudes toward alternative sources of authority. The problem here is the kind of story that adventure generates and the claim to credibility that it makes, a credibility that depends on the construction of a heroic self.

This collection brings together fieldwork and adventure because they are both presumed to be heroic. They both involve venturing into novel situations that will test one's ability to survive. By stressing that adventure is a deliberate undertaking that requires conscious choice and awareness of risk, I am departing from Simmel's definition of adventure. Simmel regards adventure as an effect of accidental circumstances, a temporary lack of control that shakes one's sense of confidence, thus offering a striking contrast to everyday life and generating a powerful sense of meaning. In my opinion, he is actually describing the anxiety-discharging, gratitude-building effects of surviving a calamity. "Thank god we're alive!" Calamity is a necessary feature of adventure—no adventure fills the bill without it. But the experiences being explored in this collection are a larger and more deliberate endeavor than mere calamity.

If adventure were simply the result of accident, we could not embark on an adventure or go in search of it—customary usages in English—because adventure could only befall one, not be actively chosen. Adventures that are more than sudden calamities imply considerable effort to achieve a certain kind of experience. The search is similar to the quest, the pilgrimage, or the mission but is more open-ended than a journey fixed on a lost grail, ark, or tribe. Adventure is a deliberate undertaking that requires daring decisions, in which the adventurer chooses to enter into situations where he will have less control than usual, where he risks losing control completely and suffering dire consequences that could have been avoided.

This makes some kinds of fieldwork a subset of adventure, but the decision-making that goes into each is intended to produce very different narratives. As scientific research, fieldwork is supposed to be planned with care, be carried out with diligence, and produce carefully considered conclusions. Above all it cannot harm informants or collaborators—this according to the American Anthropological Association's Code of Ethics. In contrast, going in search of adventure requires actively courting danger, to the point of inviting it into your canoe. Adventure minus any danger does not add up to adventure. It requires daring or foolish decisions whose unfortunate outcomes become the pivot of the story. Egregious mistakes and life-threatening mishaps are proudly owned because, without them, it would not be an adventure. The result is a narrative of survival that claims a very different kind of credibility than scientific research.

I speak as one who is presumed to have had adventures because I did fieldwork in a counterinsurgency zone in Guatemala. The truth is, I correctly calculated the risks as lower than they appeared to be, I cleverly followed in the path of aid workers who had gone before me, I figured out how to keep a low profile and, as a matter of principle, I managed to avoid having adventures with unfortunate outcomes. It is good that I have no exciting stories to tell because, among my advisers at Stanford University, there was frank skepticism that a counterinsurgency zone in Guatemala was sufficiently low-risk for fieldwork in 1988–89. They were more worried about my local informants than me. What if the Guatemalan army confiscated my notes and targeted the people who were helping me?

Of three other graduate students who did fieldwork in the western highlands during this period—Robert Carlsen, Linda Green and Judith Zur—all three managed to get through most of their planned stay like I did, but each of them had more trouble with the Guatemalan army than I. In the case of Carlsen and Zur, unwanted attention from the army forced them to end their research earlier than planned. A year later, in November 1990, army operatives murdered our Guatemalan colleague Myrna Mack. Mack was interviewing internal refugees whom the Guatemalan army regarded as a logistical base for the Guerrilla Army of the Poor. Unlike myself and the three others, Mack did not have a passport from a powerful, influential country. Also unlike the rest of us, she was involved in public challenges to the army's claims, through the Catholic Church's advocacy for the refugees in question, the Communities of Population in Resistance. This indeed was an adventure, which culminated in being knifed to death outside her office by army plainclothesmen working for the Presidential Military Staff (REMHI 1998: 292–93).

If I had experienced a serious adventure—such as getting arrested and causing the detention or disappearance of one of my sources—Stanford might well have refused to credential me as a PhD, and with good reason. Only a few years before, the Stanford Anthropology Department disciplined two graduate students for taking risks that it decided were unethical. In

1981, Steven Mosher denounced the Chinese government's enforcement of the one-child population policy by forcing women to have late-term abortions. His magazine exposé included photos showing the faces of his subjects, which made them identifiable and which Mosher himself has admitted was a mistake. The Chinese government expelled him and so did Stanford— although Stanford rested its case on Mosher's lack of candor with his committee. The same year that Mosher's criticism of the Chinese government put him under a cloud, in 1981, another Stanford graduate student named Philippe Bourgois took a sudden opportunity to visit a guerrilla-controlled zone in El Salvador. As soon as he arrived, unfortunately, the Salvadoran army attacked. Troops and helicopter gunships pursued fleeing civilians for days, with Bourgois himself surviving only because he was young and fleet. Like Mosher, Bourgois took his experiences to the media. Mosher never obtained his degree; Bourgois did; but in both cases, according to Stanford, they had not informed their committees about important risks they were taking. In both cases (although not very plausibly in Bourgois's case as far as I can see), Stanford decided that they had endangered their human subjects (Beyers 1985; Coughlin 1987; Bourgois 1990).

I now hasten to add that pulling Mack, Mosher, and Bourgois into an essay on adventure is a misrepresentation of what they were about, which I do only to make the following point. Each made the decisions that they did for what they felt were compelling reasons, not for kicks. Mosher wanted to embarrass the Chinese government into ending coerced abortions. Mack and Bourgois wanted to stop military bombardments of civilians. I certainly would not want my own research in a Guatemalan war zone to be described as an adventure. Why not? For the simple reason that counterinsurgency zones and coerced late-term abortions are life-and-death situations for the people who are trapped in them. Anyone who chooses to join these situations, then labels them as an adventure, is playing up his own bravado at the expense of others who have no choice.

Because the very term adventure underlines the cruel contrast between the haves and have-nots of this world, the people with passports and people without them, this ever more embarrassing gap has given anthropologists a professional interest in avoiding the term. Yet there is no denying that some anthropologists are drawn to risky situations, that our association with these can become an important source of professional capital, and that the entire profession has capitalized on the aura of adventure surrounding fieldwork.

Let me now distinguish two kinds of adventure which, when they occur in fieldwork, pose ethical dilemmas:

One is the calamity, the unforeseen peril that jumps out in front of you like an avalanche. Or like a zealous new lieutenant who takes a sudden interest in your research and assigns a couple of soldiers to follow you around town, causing you to panic that you, or even worse, the Guatemalans who have been helping you, will be arrested and interrogated. This kind of adventure is best described as a calamity because, while your decision-making placed you in harm's way, you were taken by surprise. You may have been vaguely aware of venturing onto dangerous ground but failed to plan for a dangerous contingency.

The second kind of adventure is a risk or challenge that you choose to advance your research, your career or personal development. Thus you might decide to approach the local army commander, in the hope of being able to watch how he uses his power vis-á-vis the local civilian authority. However, you realize that it will be risky because you will have to explain the purposes of your study. It could backfire by drawing unwelcome attention to yourself and the vulnerable local people who have been helping you.

Generally speaking, the adventure I was trying to avoid in Guatemala was a collision with a power structure. Cartoon anthropologists face adventure in the form of demanding topography or hostile natives, but what made my fieldwork adventurous was a military occupation in which I could trigger reprisals against myself or my informants. This was what worried my Stanford advisers: they couched the risk in terms of human subjects and we negotiated how I would deal with it in terms of my human subjects protocol. The protection of human subjects is a form of institutional supervision that originated in biomedical research and has now been transferred into social research, to the growing dismay of anthropologists.

In the U.S. human subjects monitoring began with the outcry over a syphilis experiment conducted by the U.S. Public Health Service at the Tuskegee Institute in Alabama. From 1932 to 1972, government researchers withheld treatment from 399 black men in advanced stages of syphilis in order to study the long-term effects (Jones 1981). They did so without their subjects' knowledge or consent, in disregard for the Hippocratic Oath, and even prevented their subjects from being treated by other doctors. So the impetus for the new regulations was to protect human guinea pigs from unethical medical experiments. Ever since, medical researchers have been required to explain the purpose of their research to their subjects, to inform them of any risks they may incur, and to obtain their informed consent, all of which is to be certified with a signed consent form.

Since the 1980s, the federal government has extended human subjects requirements to social research including participant-observation as well as interviewing. The controls no longer apply merely to federally financed research. Any institution that receives federal funding must now set up an institutional review board (IRB) to supervise all research conducted by the institution. Thus IRBs now routinely require any researcher interviewing people one-on-one or in small groups to obtain signed consent forms from each human subject. Many social researchers question whether the biomedical model is appropriate for oral communication between researchers and their sources, let alone for the more informal styles of interactional research. For anthropologists like myself who work with preliterate people, IRBs typically waive the requirement for signed consent. But signed consent forms are unwieldy for many kinds of participant-observation research—for example, with any shifting social group at a party or on a street corner. Most risks in this kind of research are slight, they are usually created by many factors beyond the researcher's control, and they are typically impossible to predict. Yet a few cases of institutions losing all their federal funding, owing to regulatory overreactions, have prompted IRBs to start rejecting social research on the basis of improbable worst-case scenarios (Overbey 2001; Bruner 2004).

In the U.S., human-subjects regulation is driven by lawyers and fear of torts, but there are parallel developments in other Western countries that Marilyn Strathern (2000) and her colleagues have analyzed in terms of "audit culture." Audit culture is not just a set of verification rituals, or a politically neutral legal-administrative practice but, in Foucaultian terms, an instrument of new forms of governance and power (Shore and Wright 2000: 57). The rationale for many new expressions of audit culture is what Michalis Lianos and Mary Douglas (2000: 267) call "dangerization," that is, "the tendency to perceive and analyse the world through categories of menace."

Dangerization has gone on display in the multiplication of video monitors and other automated security devices to surveil public areas and to mark off private from public spaces. It can also be found in the projection of the biomedical human-subjects model far beyond the medical and psychological experiments for which it was designed. The open-ended, unpredictable quality of ethnographic research has become all too attractive to human-subjects regulators. Once dangerization becomes part of an institutional routine, Lianos and Douglas (2000: 273) warn, it produces a constant "scanning [of] the environment for perceptual indices of irregularity, which are then perceived as menacing."

Once risk has been magnified, meticulous routine is required to protect against it, overriding the judgement of the person doing the research. Two kinds of fieldwork could become impossible if the biomedical model for ethical research is carried to the logical but far-fetched worst-case scenario. The first kind of research is on illegal or dangerous behavior. For example, warfare and its substitutes—such as football hooliganism—are inherently risky whether or not a researcher is present. Once worst-case logic is applied, the presence of a researcher becomes even more risky. Just imagine how a hormonally attractive researcher could encourage belligerents to demonstrate their manhood even more strenuously than before, making the research unacceptably risky. As for the informed-consent requirement, imagine calling the adversaries together, asking them to put down their weapons or gear, delivering a human subjects talk, pointing out that the presence of the researcher could increase the risks to which they are subjecting themselves, and persuading every last one to sign a legal document.

A second kind of investigation that could be regulated out of existence by the biomedical model for ethical fieldwork is research on powerholders—like those Guatemalan army officers whom I usually avoided interviewing. Laura Nader's (1969) idea of "studying up" in the power structure rather than "studying down" is one of the most valuable suggestions in the history of anthropology, but it could succumb to rigorous application of the biomedical ethical model. To echo a point made by Philippe Bourgois (1990), if my study had focused on army officers rather than on violence survivors, would I be obliged to inform an army officer that one of the risks of helping me is that his words could be turned against him? If I tape-recorded an interview in which he employed euphemisms for torturing and killing prisoners, would I be required to respect his anonymity? This is what biomedical research ethics could require. Yet to do so would violate my ethical commitment to other people I was studying—the population whom the officer was administering.

The biomedical definition of a human subject is too narrow for social research because it presumes that the subject is a powerless doctor's patient. The premise of a human subjects protocol is that the researcher has power over the subject. Yet some subjects have power over researchers, to say nothing of power over other research subjects, whose human rights they may be violating. This brings us back to the problem of adventure in fieldwork, of any kind that generates animosity toward the researcher or his collaborators. Since any kind of questioning of power figures can potentially lead to reprisals, IRBs could use the biomedical model of research ethics, and how it defines research subjects as powerless patients, to rule out research on power structures. The reason is that research on the powerful is too risky.

The reductio ad absurdum biomedical regulation I've been describing comes from outside anthropology. But the problem of how to protect informants in lethal political environments is no laughing matter. It will not go away; it requires the most careful attention by everyone in anthropology who deals with these situations. Meanwhile, there is a deeper sense in which anthropology is regulating itself against adventure, an epistemological distrust of the heroic self. This is not just a fad confined to postmodernists—it reflects a much wider lack of confidence. Elsewhere in the collection, David Napier describes it in terms of American culture but there are probably parallels elsewhere. At least in the U.S., parental protection of the single progeny, institutional fear of litigation, and consumer gratification have combined to undermine rites of passage from one life-stage to another.

Take academic life. There used to be plenty of room for pranks, drinking contests, and sexual adventure, or so the old-timers tell us. Now these are all risk behaviors. If you feel like indulging in any of them, your therapist will hold your hand and your insurance plan may even pay for a few sessions. But don't let your risk behaviors come to the attention of the human subjects committee or the diversity dean. Even in the supposedly rugged outpost of anthropology, the anthropologist in his tent has been displaced by the theorist in the armchair. Sanctimony reigns. Bug bites, dysentery, and blisters have been replaced by the stress of "writing"—laboriously reworking one's narrative through fingers on a keyboard.

In the wider consumer culture, among the more daring, rites of passage are being replaced by the optional personal adventure. Some of these experiences are so genuine that they kill the protagonist—you can read about them at the Darwin Awards, a website honoring people who kill themselves through their own stupidity. Because natural selection has removed their genes from the gene pool, they have contributed to human evolution. More often, of course, prepackaged adventures minimize the possibility of death with safety procedures.

Adventure in this sense functions as an individualized rite of passage. The adventurer has proved his fiber and achieved self-validation. He is a survivor, a veteran, or an old hand even if he had to pay for the test with a credit card. The resemblance of adventure tourism to anthropology is not just coincidental, as other contributors to this collection have pointed out. In anthropology, the high-risk decision to join a profession for which there is so little market demand is just the beginning of a lengthy and highly individualized rite of passage. But this is not a rite of passage that produces a muscular hero or explorer, at least not anymore. No longer can it produce the anthropologist as oracle like Margaret Mead—imagine one of our postcolonial divas allowing herself to be photographed in a grass skirt. Instead, the only kind of hero who is permissible in contemporary anthropology is the humble sage.

Consider Karen McCarthy Brown's account of her experiences with Vodoun healing in *Mama Lola*—this is the rite of passage as a religiously imbued growth experience. Or Orin Starn's account of fieldwork with peasant watch-committees in *Nightwatch,* which is more ironic and self-deprecating. Sophisticated ethnographers put themselves on center stage, but only for carefully measured intervals, to protect themselves against accusations of narcissism. When facing dangerous situations, they protect themselves against accusations of heroism by stressing that they were terrified. Anthropology becomes a lesson in humility.

What about the old-fashioned kind of adventure, in which the hero prevails against bandits, warlords, wild animals, or a typhoon? This would require a heroic narrative with the anthropologist, or at least a trusty sidekick, overcoming the adversary. But if the adversary is human, overcoming him violates his rights as a human subject, which has become unethical. Consider the case of Napoleon Chagnon. As suggested by Harald Prins, in one of the panels that led to this volume, the adventure tone of Chagnon's early writing spelled his doom in contemporary anthropology. Once Patrick Tierney dramatized the high mortality that the Yanomamo suffered, Chagnon's attitude seemed callous and unethical, quite apart from the many issues that need to be debated.

Contemporary anthropology is hostile to adventure because adventure is a way of investing oneself with sanctity, in the sense of certainty or unquestionability used by Roy Rappaport (1999) in his cybernetics of sanctity. Usually we think of unquestionability as a quality of religious thought, but it also can be found in political and scholarly discourse, and it also can be discerned in the aura surrounding any credible survivor narrative, which is what adventure stories are—survivor narratives. When adventurers tell how they survived death and destruction, they are invoking the old association between death and unquestionability. The association between death and unquestionability is expedited by the fact that death is a definitive state, defined by the absence of life, that can be used to validate the kind of binary communication that we associate with yes/no, either/or, and surrender-or-die. This is why criminal syndicates and death squads leave bodies in public places, to send an unequivocal message. In other contexts as well, death is a subject that we all care about, that commands attention, and that is more likely to be taken seriously than any other. As an end-point that we all face, it is inherently significant. The invocation of death—or of a near-scrape with death, which is the point of a good adventure story—conveys a sense of unquestionability for which human beings hanker.

But this sort of maneuver cannot go unquestioned in a discipline dedicated to questioning Western power and epistemology. The only ethically permissible source of sanctimony is now identification with victims. Adventure is a heroic undertaking and anthropology has no room for heroics that require vanquishing a human adversary. What is left for anthropologists is education in humility, which is why the only voice left in which adventure can be performed is that of irony. This is a new form of regulation, far more internalized than human subjects regulation, and it is becoming hard to imagine doing anthropology any other way.

Bibliography

Ackerman, Robert. 1987. *J.G. Frazer: His Life and Work*. Cambridge: Cambridge University Press.

Adams, Jonathan and Thomas McShane. 1992. *The Myth of Wild Africa: Conservation Without Illusion*. Berkeley: University of California Press.

Adler, Judith. 1989. "Travel as Performed Art." *American Journal of Sociology* 94: 1366–91.

Adler, Margot. 1979. *Drawing Down the Moon: Witches, Druids, Goddess-Worshippers, and other Pagans in America Today*. Boston: Beacon.

Agamben, Giorgio. 2005. *State of Exception*. Trans. Kevin Attell. University of Chicago Press.

Aldiss, Brian W. 1986. *Trillion Year Spree: The History of Science Fiction*. London: Victor Gollancz.

Anderson, Benedict. 1991. *Imagined Communities*. London: Verso.

Andrew, Rick. 2001. *Buried in the Sky*. Sandton: Penguin.

Anon. 1985. "Ends of the Rifle." *Frontline*. August: 54.

Anon. 1990a. *Untag in Namibia*. Windhoek: Untag.

Anon. 1990b. *Namibia En Route*. Bern: Eda.

Anon. 1992. "Excerpts From Pentagon's Plan: 'Prevent Re-Emergence of a New Rival'." *New York Times*, March 7.

Apter, Michael. 1992. *The Dangerous Edge*. New York: Free Press.

Archer, Richard. 1995. "The Catholic Church in East Timor." In *East Timor at the Crossroads: The Forging of a Nation*, eds. Peter Carey and G. Carter Bentley. Honolulu: University of Hawai'i Press.

Arewa, Olufunmilayo Bamidele. 1988. *Tarzan, primus inter primates: Difference and hierarchy in popular culture*. Ph.D. Dissertation. University of California at Berkeley.

Armstrong, David. 2002. "Dick Cheney's Song of America: Drafting a Plan for Global Dominance." *Harper's Magazine*. October. 305. No. 1829:76–84.

Auerbach, Edward. 1953. *Mimesis: The Representation of Reality in Western Literature*, trans. William Trask. Princeton: Princeton University Press.

Auster, Paul. 2002. *Book of Illusions*. London: Faber and Faber.

Auzepy, Marie-France et Joel Cornette. 2003. *Palais et pouvoir: de Constantinople a Versailles*. Vincennes: Presses universitaires de Vincennes.

Axelrod, Charles David. 1979. *Studies in Intellectual Breakthrough: Freud, Simmel, Buber*. Amherst: University of Massachusetts Press.

Bachofen, Johannes J. 1861. *Das Mutterrecht: Eine Untersuchung über die Gynaikokratie der alten Welt nach ihrer religiösen und rechtlichen Natur*. Stuttgart: Krais und Hoffman.

Baines, Gary. 2002. "'South Africa's Vietnam'? Literary History and Cultural Memory of the Border War." In *Telling Wounds*. Proceedings of the conference held at the University of Cape Town, 3–5 July 2002, eds. Chris van der Merwe and Rolf Wolfswinkel.

Barnard, Alan. 1989. "The Lost World of Laurens van der Post?" *Current Anthropology* 30:104–14.

———— 1994. "Tarzan and the Lost Races: Parallels between Anthropology and Early Science Fiction." In *Exploring the Written: Anthropology and the Multiplicity of Writing*, ed. Eduardo P. Archetti. Oslo: Scandinavian University Press.

———— 1995a. "*Orang Outang* and the Definition of *Man*: The Legacy of Lord Monboddo." In *Fieldwork and Footnotes: Studies in the History of European Anthropology*, eds. Han F. Vermeulen and Arturo Alvarez Roldan. London and New York: Routledge.

———— 1995b. "Monboddo's Orang Outang and the Definition of Man." In *Ape, Man, Apeman: Changing Views since 1600*. Eds. Raymond Corbey and Bert Theunissen. Leiden: Department of Prehistory.

———— 2000. *History and Theory in Anthropology*. Cambridge: Cambridge University Press.

Barringer, Tim. 1996. "Fabricating Africa: Livingstone and the Visual Image 1850–1874." In *David Livingstone and the Victorian Encounter with Africa*. London: National Portrait Gallery.

Bauman, Zygmunt. 1997. *Postmodernity and its Discontents*. New York: New York University Press.

———— 2002. *Society Under Siege*. Malden, MA: Polity Press.

Bayers, Peter. 2003. *Imperial Ascent: Mountaineering, Masculinity, and Empire*. Boulder: University Press of Colorado.

Beard, Mary. 1992. "Frazer, Leach, and Virgil: The Popularity (and Unpopularity) of *The Golden Bough*." *Comparative Studies in Society and History* 34:203–24.

Beedie, Paul. 2003. "Adventure Tourism." In *Sport and Adventure Tourism,* ed. Simon Hudson. Binghamton: Haworth.

Beedie, Paul, and Simon Hudson. 2003. "Emergence of Mountain-Based Adventure Tourism." *Annals of Tourism Research* 30(3):625–43.

Bendix, Regina. 1997. *In Search of Authenticity.* Madison: University of Wisconsin Press.

Benjamin, Walter. 1969. "The Work of Art in the Age of Mechanical Reproduction." In *Illuminations,* ed. Hannah Arendt. New York: Schocken.

———— 1999. *The Arcades Project,* trans. Howard Eiland and Kevin McLaughlin. Cambridge: Belknap Press.

Bennett Ross, Jane. 1984. "Effects of Contact on Revenge Hostilities Among the Achuara Jívaro." In *Warfare Culture, and Environment,* ed. R.B. Ferguson Orlando: Academic Press.

Berger, John. 1980. "Why look at Animals?" *About Looking.* New York: Pantheon.

Berman, Marshall. 1982. *All That is Solid Melts into Air: The Experience of Modernity.* New York: Simon and Schuster.

Bernstein, Peter L. 1996. *Against the Gods: The Remarkable Story of Risk.* New York: John Wiley.

Beyers, Bob. 1985. "Kennedy upholds Mosher's termination." *The Stanford University Campus Report.* October 2:1, 9–19.

Bigly, Cantell A. [pseud. of George Washington Peck]. 1849. *Aurifodina; or, Adventures in the Gold Region.* New York: Baker and Scribner.

Bloom, Lisa. 1993. *Gender on Ice: American Ideologies of Polar Expeditions.* Minneapolis: University of Minnesota Press.

Bodry-Saunders, Susan. 1991. *Carl Akeley: Africa's Collector, Africa's Savior.* New York: Paragon House.

Bonington, Chris. 1991. *Quest for Adventure.* London: Hodder and Stoughton.

Boorstin, Daniel. 1963. *The Image: A Guide to Pseudo-Events in America.* New York: Vintage.

Bourdieu, Pierre. 1998. "The essence of neo-liberalism." *Le Monde Diplomatique* 12.

Bourgois, Philippe. 1990. "Confronting Anthropological Ethics: Ethnographic Lessons from Central America." *Journal of Peace Research* 27(1):43–54.

Bourke, Joanna. 1999. *An Intimate History of Killing.* New York: Basic.

Bousé, Derek. 2000. *Wildlife Films.* Philadelphia: University of Pennsylvania Press.

Bowman, Glenn. 1989. "Fucking Tourists: Sexual Relations and Tourism in Jerusalem's Old City." *Critique of Anthropology.* 9(2):77–93.

Boyle, Andrew. 1977. *The Riddle of Erskine Childers.* London: Hutchinson.

Bradburd, Daniel. 1998. *Being There: The Necessity of Fieldwork.* Washington D.C.: Smithsonian Institution Press.

_____ 2000a . "Instrumental encounters: Tribal Iran in Travel Writing and Ethnography," presented at the Biennial Conference of the European Association of Social Anthropologists, Krakow (Poland), 26–29 July 2000.

_____ 2000b. "A Burning Desire To Consume: Red Hot Food In Contemporary America." Presented at the 99th Annual Meeting of the American Anthropological Association, San Francisco, CA.

_____ 2001. "Snow Mobiles And Snow Shoes: Alternative Responses To The Crisis Of Modernity." Presented at the 100th Annual Meeting of the American Anthropological Association, Washington, D.C.

Brantlinger, Patrick. 1988. *Rule of Darkness*. Ithaca: Cornell University Press.

Bright, Brenda Jo and Liza Bakewell, eds. 1995. *Looking High and Low: Art and Cultural Identity*. Tucson: University of Arizona Press.

Brouwer, H.A. ed. 1925. *Practical Hints to Scientific Travelers*. (2 Volumes) 2nd Revised Edition. The Hague: Martinus Nijhoff.

Browder, Laura. 2000. *Slippery Characters: Ethnic Impersonators and American Identities*. Chapel Hill: University of North Carolina Press.

Brown, Karen McCarthy. 1990. *Mama Lola*. Berkeley: University of California Press.

Brown, Michael F. 1985. *Tsewa's Gift: Magic and Meaning in an Amazonian Society*. Washington: Smithsonian Institution Press.

Brown, Robert. 1986. *Soldier of Fortune*. New York: Exeter House.

Bruner, Edward. 1986. "Experience and Its Expressions." In *The Anthropology of Experience*, eds. Edward Bruner and Victor Turner. Urbana: University of Illinois Press.

——— 1995. "The Ethnographer/Tourist in Indonesia." In *International Tourism: Identity and Change*, eds. Lanfant, Marie-Françoise, et al. Thousand Oaks, California and London: Sage Publications.

——— 2004. "Ethnographic Practice and Human Subjects Review." *Anthropology News* 45(1): 10, January.

Bruno, Paola. 2003. "Una chica de 14 años, la más joven en hacer cumbre en el Aconcagua." *Los Andes* 7 Feb 2003. < http://www.losandes.com.ar/nrc=110630>.

Bryson, Bill. 1989. *The Lost Continent*. New York: Harper & Row.

Bull, Bartle. 1992. *Safari*. New York: Penguin.

Burdick, Eugene, and William J. Lederer. 1958. *The Ugly American*. New York: Norton.

Burroughs, Edgar Rice. 1914. *Tarzan of the Apes*. New York: A.L. Burt.

——— 1933 [1931/1932]. *Jungle Girl* [The Land of Hidden Men]. London: Odhams Press.

——— 1963a [1912/1914]. *Tarzan of the Apes* (Tarzan Series 1). New York: Ballantine Books.

——— 1963b [1913/1915]. *The Return of Tarzan* (Tarzan Series 2). New York: Ballantine Books.

—— 1963c [1913–17/1925]. *The Cave Girl*. New York: Ace Books.

—— 1963d [1914/1922]. *At the Earth's Core* (Pellucidar Series 1). New York: Ace Books.

—— 1963e [1914–15/1925]. *The Eternal Savage* [The Eternal Lover]. New York: Ace Books.

—— 1963f [1916]. *The Lost Continent* [Beyond Thirty]. New York: Ace Books.

—— 1963g [1916/1918]. *Tarzan and the Jewels of Opar* (Tarzan Series 5). New York: Ballantine Books.

—— 1963h [1917/1938]. *The Lad and the Lion*. New York: Ace Books.

—— 1975 [1918/1924]. *The Land That Time Forgot*. London: Tandem.

—— 1983 (orig. 1912). *Tarzan of the Apes*. New York: Ballantine.

Bush, President George. 2004. "State of the Union Address." *New York Times*. 21 January. p. 1.

Campbell, Martin, dir. 2001. *Vertical Limit*. Columbia Pictures.

Camus, Albert. 1962. *Carnets 1935–37*. Paris: NRF/Gallimard.

Canetti, Elias. 1962 (1960). *Crowds and Power*, trans. Carol Stewart. London: Victor Gollancz Ltd.

Caplan, Patricia, ed. 2000. *Risk Revisited*. London: Pluto.

Carlsen, Robert S. 1997. *The War for the Heart and Soul of a Highland Maya Town*. Austin: University of Texas Press.

Carton, Benedict. 2000. "Unfinished Exorcism: the Legacy of Apartheid in Democratic Southern Africa." *Social Justice* 27:116–27.

Castel, Robert. 1991. "From Dangerousness to Risk." In *The Foucault Effect: Studies in Governmentality*, eds. Graham Burchell, Colin Gordon and Peter Miller. Chicago: University of Chicago Press.

Chalk, Peter. 2001. *Australian Foreign and Defense Policy in the Wake of 1999/2000 East Timor Intervention*. Santa Monica, CA: Rand Corporation.

Childers, Erskine. 1995 [1903]. *The Riddle of the Sands*. London: Oxford University Press.

Choudry, Aziz. 2003. "Tarzan, Indiana Jones and Conservation International's Global Greenwash Machine." WWW site, accessed on 12 November 2003. http://www.organicconsumers.org/chiapas/conservation_international.cfm

Clareson, Thomas D. 1977 [1975]. "Lost Lands, Lost Races: A Pagan Princess of Their Very Own." In *Many Futures, Many Worlds: Theme and Form in Science Fiction*. Kent, Ohio: Kent State University Press.

—— 1984. *Science Fiction in America, 1870s–1930s: An Annotated Bibliography of Primary Sources*. Westport, CT: Greenwood Press.

—— 1985. *Some Kind of Paradise: The Emergence of American Science Fiction*. Westport, CT: Greenwood Press.

Clark, Leonard. 1953. *The Rivers Ran East*. New York: Funk and Wagnalls.

Clark, Steve, ed. 1999. *Travel Writing & Empire: Postcolonial Theory in Transit.* London and New York: Zed Books.

Clastres, Pierre. 1989. *Society Against the State.* New York: Zone Books.

Clifford, James. 1997. *Routes: Travel and Translation in the Late Twentieth Century.* Cambridge and London: Harvard University Press.

Cloke, Paul, and Harvey C. Perkins. 1998. "Cracking the canyon with the awesome foursome: representations of adventure tourism in New Zealand." *Environment and Planning D: Society and Space* 16:185–218.

Clute, John, and Peter Nicholls, eds. 1993. *The Encyclopedia of Science Fiction.* New York: St. Martins.

Coburn, Broughton. 1997. *Everest : Mountain without Mercy.* Washington, DC: National Geographic Society.

Cohen, Stan, and Laurie Taylor. 1992. *Escape Attempts.* New York: Routledge.

Cohodas, Marvin. 1997. *Basket Weavers for the California Curio Trade: Elizabeth and Louise Hickox.* Tucson: University of Arizona Press.

———— 1999. "Elizabeth Hickox and Karuk Basketry." In *Unpacking Culture, Art and Commodity in Colonial and Post Colonial Worlds,* eds. Ruth Phillips and Christopher Steiner. Berkeley: University of California Press.

Colletta, Nat, Martin Kostner and Ingo Wiederhofer. 1996. *The Transition from War to Peace in Sub-Saharan Africa.* Washington DC: World Bank.

Colls, Rachel. 2003. "Tourism: Between place and performance by Coleman, S, Grang M." *Social and Cultural Geography* 4(1):119–21.

Connor, Steven. 1978. *Postmodernist Culture.* Oxford: Blackwell.

Conrad, Joseph. 1988 [1899]. *Heart of Darkness.* New York: Norton.

Coser, Lewis A. 1965. *Georg Simmel.* Englewood Cliffs, N.J.: Prentice-Hall.

———— 1977. *Masters of Sociological Thought: Ideas in Historical and Social Context.* 2nd ed. New York: Harcourt Brace Jovanovich.

Coughlin, Ellen K. 1987. "Politics and Scholarship Mix in China Researcher's Long Battle with Stanford." *Chronicle of Higher Education,* 33(18):1,8–10, January 14.

Courtwright, David. 1998. *Violent Land: Single Men and Social Disorder from the Frontier to the Inner City.* Cambridge: Harvard University Press.

Craik, Jennifer. 1997. "The Culture of Tourism." In *Touring Cultures: Transformation of Travel and Theory,* eds. Chris Rojek and John Urry. London: Routledge.

Crawshaw, Carol, and John Urry. 1997. "Tourism and the Photographic Eye." In *Touring Cultures: Transformations of Travel and Theory,* eds. Chris Rojek and John Urry. London and New York: Routledge.

Crouch, Gregory. 2002. *Enduring Patagonia.* New York: Random House.

Dabbs, James M. 2000. *Heroes, Rogues and Lovers: Testosterone and Behavior.* New York: McGraw-Hill.

Darwin, Charles. 1871. *The Descent of Man and Selection in Relation to Sex.* London: John Murray.

Davis, Murray S. 1973. "Georg Simmel and the Aesthetics of Social Reality." *Social Forces* 51(3):320–29.

Davis, Susan. 1997. *Spectacular Nature: Corporate Culture and the Sea World Experience.* Berkeley: University of California Press.

Day, Beth. 1955. *America's First Cowgirl Lucille Mulhall.* New York: Julian Messner.

DeBord, Guy. 1994. *The Society of the Spectacle,* trans. Donald Nicholson-Smith. New York: Zone Books.

Dening, Greg. 1994. "The theatricality of observing and being observed: Eighteenth-century Europe "discovers" the ? century 'Pacific'." In *Implicit Understandings,* ed. Stuart B. Schwartz. New York: Cambridge University Press.

Denzin, Norman. 1989. *Interpretive Interactionism.* Newbury Park: Sage.

Descola, Philippe. 1994. *In the Society of Nature,* trans. N. Scott. Cambridge: Cambridge University Press.

——— 1996a. *The Spears of Twilight,* trans. Janet Lloyd. New York: The New Press.

——— 1996b. "Constructing Natures: Symbolic Ecology and Social Practice," in *Nature and Society,* eds. Philippe Descola and Gíseli Pálsson. New York: Routledge.

Dolan, Brian. 2001. *Ladies of the Grand Tour: British Women in Pursuit of Enlightenment and Adventure in Eighteenth-Century Europe.* New York: HarperCollins.

Donnelly, Thomas. 2000. *Rebuilding America's Defenses: Strategy, Forces and Resources For A New Century.* Washington DC: Report of The Project for the New American Century.

Douglas, Mary, and Aaron Wildavsky. 1982. *Risk and Culture.* Berkeley: University of California Press.

Dowie, Michael. 1995. *Losing Ground: American Environmentalism at the close of the Twentieth Century.* Cambridge: MIT Press.

Doyle, Arthur Conan. 1936 [1912]. *The Lost World.* London: Hodder and Stoughton.

Dunn, Kevin. 1996. "Lights ... Cameras ... Africa: Images of Africa and Africans in Western Popular Films." *African Studies Review* 39(1):149–76.

Dunn, Molly. 2004. "Paradox of Agency: Confidence and Fatalism in Peace Corps Volunteers." B.A. thesis. New York University.

Durkheim, Emile. 1965. *The Elementary Forms of Religious Life.* New York: The Free Press.

Eberhardt, Isabelle. 1991. *The Oblivion Seekers.* San Francisco: City Lights.

Ebron, Paulla. 1997. "Traffic in Men." *Gendered Encounters: Challenging Cultural Boundaries and Social Hierarchies in Africa,* eds. M. Grosz-Ngate and O.H. Kokole. London: Routledge.

Elizondo, Virigilio P. and Sean Freyne, eds. 1996. *Pilgrimage.* London: SCM Press.

Enloe, Cynthia. 1999. *Maneuvers.* Berkeley: University of California Press.

Enstad, Nan. 1999. *Ladies of Labor, Girls of Adventure: Working Women, Popular Culture, and Labor Politics at the Turn of the Twentieth Century.* New York: Columbia University Press.

Enzensberger, Hans Magnus. 1997. "A Theory of Tourism." *New German Critique* 68:117–35.

Escobar, Arturo. 1995. *Encountering Development: The Making and Unmaking of the Third World.* Princeton: Princeton University Press.

Essoe, Gabe. 1968. *Tarzan of the Movies: A Pictorial History of More Than Fifty Years of Edgar Rice Burroughs' Legendary Hero.* New York: Cadillac Publishing Co.

Exodus. 2003/2004. "Real traveler, exploring the real world." *Overland journeys travel brochure.*

Explorers Club. 1941. *Through Hell and High Water.* New York: Robt. McBride & Co.

Fabian, Johannes. 2000. *Out of Our Minds: Reason and Madness in the Exploration of Central Africa.* Berkeley: University of California Press.

Fausto, Carlos. 2000. "Of Enemies and Pets: Warfare and Shamanism in Amazonia." *American Ethnologist* 26(4):933–56.

Fedarko, Kevin. 2003. "The Mountain of Mountains." *Outside.* November. 27:97+.

Feinstein, Anthony. 1998. *In Conflict.* Windhoek: New Namibia Books.

Ferguson, James. 1992. *The Anti-Politics Machine.* Cape Town: D. Philip.

Firth, Raymond. 1957. "Introduction: Malinowski as Scientist and as Man." *Man and Culture: An Evaluation of the Work of Bronislaw Malinowski.* Ed. Raymond Firth. London: Routledge & Kegan Paul.

Fischer, Fritz. 1998. *Making Them Like Us: Peace Corps Volunteers in the 1960s.* Washington, DC: Smithsonian Institution Press.

Fitz Gerald, Edward A. 1899. *The Highest Andes.* New York: Scribner.

Fitzgerald, Frances. 2002. "George Bush and the World." *New York Review of Books,* 49(14) (September 26):80–86.

Fleming, Fergus. 1998. *Barrow's Boys.* London: Granta.

Fowler, Barry. ed. 1995. *Pro Patria.* Halifax: Sentinal Projects.

Fowler, Barry. 1996. *Grensvegter?* Halifax: Sentinal Projects.

Fox, Richard W. and T.J. Jackson Lears, eds. 1983. *The Culture of Consumption: Critical Essays in American History, 1880–1980.* New York: Pantheon Books.

Foy, Jessica, and Karal Ann Marling. 1994. *The Arts and the American Home: 1890–1930.* Knoxville: University of Tennessee Press.

Franklin, Adrian. 1999. *Animals and Modern Cultures.* Oxford: Sage.

Frazer, Sir James George. 1922. *The Golden Bough: A Study in Magic and Religion* (abridged edition). London: Macmillan and Co.

Freud, Sigmund. 1961 [1930]. *Civilization and Its Discontents,* trans. J. Strachey. New York: Norton.

Frey, Nancy Louise. 1998. *Pilgrim Stories: On and Off the Road to Santiago.* Berkeley: University of California Press.

Friese, Kai. n.d. "White Skin, Black Mask." In *Transition* 80.4–17.

Frisby, David, and Mike Featherstone, eds. 1997. *Simmel on Culture: Selected Writings.* London: Sage Publications.

Fussell, Paul, ed. 1987. *The Norton Book of Travel*. New York: W.W. Norton.

G.A.P, Adventures. 2003/2004. "1 World for the New Traveler." *Small Group Adventures and Safaris travel brochure*.

Gallardy, Louis. 1953. "Introduction." *The Rivers Ran East*. New York: Funk and Wagnalls Company.

Gallegos, Romulo. 1935. *Canaima*. Barcelona: Araluce.

―――― 1984. *Canaima,* trans. Jaime Tello. Norman: University of Oklahoma Press.

Galton, Francis. 1855. *The Art of Travel, or, Shifts and Contrivances Available in Wild Countries*. London: John Murray.

―――― 1889. *Narrative of an Explorer in Tropical South Africa*. London: Ward, Lock Co.

―――― 1914. *Hereditary Genius*. London: Macmillan.

―――― 2000 [1872]. *The Art of Travel, or, Shifts and Contrivances Available in Wild Countries*. Fifth Edition. London: Phoenix Press.

García, Nicolás. 2003a. "Aconcagua, un negocio de U$S 10 millones." *Los Andes* 25 Jan. 2003. <http://www.losandes.com.ar/nrc=

―――― 2004. "Aconcagua: récord de visitas y recaudación." *Los Andes* 16 Mar. 2004. <http://www.losandes.com.ar/nrc=177019>

―――― 2003b. "Grajales: 'Es parte de nuestra cultura'." 2003 *Los Andes* 25 Jan. 2003. http://www.losandes.com.ar/asp?nrc

Gardner, Gerald. 1955 *Witchcraft Today.* New York: Citadel.

Geertz, Clifford. 1973. *The Interpretation of Cultures*. New York: Basic Books.

―――― 1988. *Works and Lives: the Anthropologist as Author.* Stanford: Stanford University Press.

Gibson, James W. 1994. *Warrior Dreams: Violence and Manhood*. New York: Hill and Wang.

Gilman, Sander. 1985. *Difference and Pathology: Stereotypes of Sexuality, Race and Madness*. Ithaca: Cornell University Press.

Gippelhauser, Richard, and Elke Mader. 1990. *Die Achuara-Jívaro: Wirtschaftliche und Soziale Organisationsformen am Peruanischen Amazonas*. Wien: Verlag der osterreichsichen Akademie der Wissenschaften.

Goffman, Erving. 1967. "Where the Action Is." *Interaction Ritual*. New York: Anchor.

Gora, Bronwen. 2002. "Ask the Icon—Steve Irwin, Crocodile Hunter and Aussie Legend." *The Sunday Telegraph* (Sydney). October 6th p. 124.

Gordon, Robert J. 1991. "Marginalia on 'Grensliteratuur': Or How/Why is Terror Culturally Constructed in Northern Namibia?" *Critical Arts* 5(3):79–95.

―――― 1997. *Picturing Bushmen*. Athens: Ohio University Press.

Goudge, Paulette. 2003. *The Whiteness of Power.* London: Lawrence and Wishart.

Goulding, Marack. 2002. *Peacemonger.* London: John Murray.

Green, Linda. 1999. *Fear as a Way of Life*. New York: Columbia University Press.

Green, Martin. 1993. *The Adventurous Male*. University Park: Pennsylvania State University Press.

Griffith, Allan. 1998. *Conflict and Resolution: Peace-building through the Ballot Box in Zimbabwe, Namibia and Cambodia*. Oxford: New Cherwell Press.

Griffiths, Alison. 2002. *Wondrous Difference: Cinema, Anthropology and Turn-of-the Century Visual Culture*. New York: Columbia University Press.

Gugler, Josef. 2003. *African Film: Re-Imagining a Continent*. Bloomington: Indiana University Press.

Guha, Ramachandra. 2000. *Environmentalism: A Global History*. New York: Longman.

Haddon, Alfred C. 1910. *History of Anthropology*. London: Watts & Co.

Haggard, H. Rider and Andrew Lang 1972 [1890]. *The World's Desire*. New York: Ballantine Books.

Hamilton, Gary G. 1978. "The Structural Sources of Adventurism: The Case of the California Gold Rush." *American Journal of Sociology* 83(6) May: 1466–1490.

Hammond, Dorothy, and Alta Jablow. 1970. *The African That Never Was: Four Centuries of British Writing about Africa*. New York: Twayne Publishers.

Hancock, Graham. 1989. *Lords of Poverty: the Power, Prestige, and Corruption of the International Aid Business*. New York: Atlantic Monthly Press.

Handler, Richard. 1986. "Authenticity." *Anthropology Today* 2(1):2–4.

Handler, Richard, and Joycelyn Linnekin. 1984. "Tradition, Genuine or Spurious." *Journal of American Folklore*, 97: 273–90.

Hansen, Anders, ed. 1993. *The Mass Media and Environmental Issues*. Leicester: Leicester University Press.

Haraway, Donna. 1989. *Primate Visions: Gender, Race, and Nature in the World of Modern Science*. New York: Routledge.

Harlech-Jones, Brian. 1997. *A New Thing?* Windhoek: EIN.

Harner, Michael. 1975. "Scarcity, the Factors of Production, and Social Evolution." *Population, Ecology, and Social Evolution*. ed. Steven Polgar. Mouton Publishers: The Hague.

———— 1984. *The Jivaro, People of the Sacred Waterfalls*. Berkeley: University of California Press.

Harris, Wilson. 1995. "Kanaima." *Concert of Voices*, ed.V.J. Ramraj. Peterborough: Broadview.

Harrison, Julia. 2003. *Being a Tourist: Finding Meaning in Pleasure Travel*. Vancouver and Toronto: University of British Columbia Press.

Hartmann, Wolfram, Jeremy Silvester, and Patricia Hayes, eds. 1999. *The Colonizing Camera*. Athens: Ohio University Press.

Harvey, David. 1987. *The Condition of Postmodernity*. Oxford: Blackwell.

Hastings, Julie. 1988. "Time out of Time: Life crises and Schooner sailing in the Pacific." *Kroeber Anthropological Papers* 67–68:42–45.

Haut, Woody. 1995. *Pulp Culture: Hardboiled Fiction and the Cold War.* New York: Serpents Tail.

Hearn, Roger. 1999. *UN Peacekeeping in Action: the Namibian Experience.* Commack, NY: Nova Science Publishers.

Hendricks, Janet Wall. 1993. *To Drink of Death.* Tucson: The University of Arizona Press.

Hepworth, Mike and Bryan S. Turner. 1982. *Confession.* New York: Routledge.

Herold, Edward, Rafael Garcia, and Tony DeMoya. 2001. Female Tourists and Beach Boys Romance or Sex Tourism? *Annals of Tourism Research* 28 (4):978–97.

Hightower, Jim. 2004. "The Smell of Hummer." Posted October 9 on AlterNet. http://www.alternet.org/columnists/story/20126/.

Hobsbawm, Eric, and Terence Ranger, eds. 1983. *The Invention of Tradition.* Cambridge: Cambridge University Press.

Hoffman, Elizabeth. 1998. *All You Need is Love: Peace Corps and the Spirit of the 1960s.* Cambridge: Harvard University Press.

Holland, Tom. 2004. "It's the flaws that make heroes so perfect." *Guardian Weekly* September 24–30:27.

Hollinshead, Keith. 1999. "Tourism as Public Culture: Horne's Ideological Commentary on the Legerdemain of Tourism." *International Journal of Tourism Research* 1:267–92.

Holmes, Burton. 1977. *The Man who Photographed the World.* New York: H. Abrams.

Holtsmark, Erling B. 1981. *Tarzan and Tradition: Classical Myth in Popular Literature.* Westport, CT: Greenwood Press.

——— 1986. *Edgar Rice Burroughs.* Boston: Twayne Publishers.

Holyfield, L., and Gary A. Fine. 1997. "Adventure as character work: The collective taming of fear." *Symbolic Interaction* 20 (4):343–63.

Hope, Christopher. 1996. *Darkest England.* Macmillan: London.

Houston, Charles S. 1998. *Going Higher: Oxygen, Man and Mountains.* Seattle: The Mountaineers.

——— 2002. "Selected Military Operations in Mountain Environments: Some Medical Aspects." In *Medical Aspects of Harsh Environments, Vol. 2.* eds. K. Pandolf and R. Burr. Washington D.C.: Borden Institute.

Houston, Charles S., and Robert H. Bates. 1979. *K2 The Savage Mountain.* Seattle: The Mountaineers.

Howard, Lise. 2002. "UN Peace Implementation in Namibia: The Causes of Success." *International Peacekeeping* 9(1):99–132.

Howes, David, ed. 1996. *Cross-Cultural Consumption: Global Markets, Local Realities.* New York: Routledge.

Hudson, Simon, ed. 2003. *Sport and Adventure Tourism.* New York: Haworth.

Huey, Richard, and Richard Salisbury (Forthcoming). "Success and Death on Mount Everest: How the Main Routes and Seasons Compare." *American Alpine Club*

Journal. Available Online http://www.americanalpineclub.org/knowledge/aaj
.asp

Hutton, Ronald. 1999. *The Triumph of the Moon*. New York: Oxford University Press.

Imperato, Pascal James, and Eleanor M. Imperato. 1992. *They Married Adventure: The Wandering Lives of Martin and Osa Johnson*. New Brunswick: Rutgers University Press.

Irwin, Steve. 2001. "Crocodiles of the Revolution." Accessed November 7th, 2001. http://www.crocodilehunter.com/crocodile_hunter/steve_travel_diary.

Irwin, Steve, and Terry Irwin. 1997. *The Crocodile Hunter: The Birthday Present was a Python and Other Adventures*. Auckland: Viking Books.

Iyer, Pico. 1986. *Video Night in Katmandu*. New York: Pantheon.

Jackson, Anthony, ed. 1986. *Anthropology at Home*. London: Tavistock Publications.

Jackson, Michael. 1998. *Minima Ethnographica: Intersubjectivity and the Anthropological Project*. Chicago: University of Chicago Press.

Jameson, Fredric. 1991. *Postmodernism or, The Cultural Logic of Late Capitalism*. Durham: Duke University Press.

Johnson, Barbara, and Ted Edwards. 1994. "The Commodification of Mountaineering." *Annals of Tourism Research* 21(3):459–78.

Johnson, Martin. 1913. *Through the South Seas with Jack London*. New York: Dodd, Mead & Co.

————— 1922. *Cannibal-Land: Adventures with a Camera in the New Hebrides*. Boston: Houghton Mifflin Company.

Johnson, Osa. 1940. *I Married Adventure: The Lives and Adventures of Martin and Osa Johnson*. Philadelphia: J. B. Lippincott Company.

————— 1944. *Bride in the Solomons*. Boston: Houghton Mifflin Company.

Jokinen, Eeva, and Soila Veijola. 1997. "The Disorientated Tourist: The Figuration of the Tourist in Contemporary Cultural Critique." In *Touring Cultures: Transformation of Travel and Theory*, eds. Chris Rojek and John Urry. London: Routledge.

Jones, Max. 2004. *The Last Great Quest: Captain Scott's Antarctic Sacrifice*. London: Oxford University Press.

Kaplan, Caren. 1998. *Questions of Travel: Postmodern Discourses of Displacement*. Durham and London: Duke University Press.

Karsten, Rafael. 1935. *The Headhunters of Western Amazonas. The Life and Culture of the Jíbaro Indians of Eastern Ecuador and Peru*. Helsinki: Societas Scientiarum Fennica, Commentationes Humanarum Littararum VII(I).

Kasson, John. 2001. *Houdini, Tarzan and the Perfect Man*. New York: Hill and Wang.

Keating, Peter. 1991 [1989]. *The Haunted Study: A Social History of the English Novel, 1875–1914*. London: Fontana Press.

Kipling, Rudyard. 1894. *The Jungle Book*. New York: Doubleday.

————— 1899. "The Tomb of His Ancestors." In *The Day's Work*. New York: Doubleday and McClure and Company.

Kipnis, Laura. 2003. *Against Love*. New York: Pantheon.

Kondo, Dorrine. 1986. "Dissolution and reconstitution of self: Implications for Anthropological Epistemology." *Cultural Anthropology* 1:74–88.

Krakauer, Jon. 1997. *Into Thin Air.* London: Pan Books.

Kuper, Adam. 1988. *The Invention of Primitive Society: Transformations of an Illusion.* London: Routledge.

—— 1996 [1973]. *Anthropology and Anthropologists: The Modern British School.* London: Routledge.

Lacassin, Francis. 1971. *Tarzan, ou le chevalier crispé.* Paris: Union General Editions.

Leach, Edmund R. 1961. "Golden Bough or Gilded Twig?" *Daedalus* 90/2:371–87.

—— 1966. "On the 'Founding Fathers.'" *Current Anthropology* 7:559–67.

—— 2000 [1965]. "On Frazer and Malinowski." In *The Essential Edmund Leach,* eds. Stephen Hugh-Jones and James Laidlaw. New Haven: Yale University Press.

Lears, T.J. Jackson. 1983. "From Salvation to Self-Realization: Advertising and the Therapeutic Roots of the Consumer Culture." In *The Culture of Consumption: Critical Essays in American History, 1880–1980,* eds. R.W. Fox and T.J. Jackson Lears. New York: Pantheon Books.

—— 1994. *Fables of Abundance.* New York: Basic Books.

Leopold, Aldo. 1949. *A Sand County Almanac.* Oxford: Oxford University Press.

Lévi-Strauss, Claude. 1962. *La Pensée sauvage.* Paris: Plon.

—— 1966. *The Savage Mind.* London: Weidenfeld & Nicolson.

—— 1973. *Tristes Tropiques,* trans. John and Doreen Weightman. London: Jonathan Cape.

Levine, Donald. 1979. "Simmel at a Distance: On the History and Systematics of the Sociology of the Stranger." In *Strangers in African Societies,* eds. William A. Shack and Elliott P. Skinner. Berkeley: University of California Press.

—— 1985. *The Flight from Ambiguity.* Chicago: University of Chicago Press.

Levine, Donald, Ellwood B. Carter, and Eleanor Miller Gorman. 1976. "Simmel's Influence on American Sociology, I." *The American Journal of Sociology* 81 (January): 813–45.

Lewis, Neil. 2000. "The Climbing Body, Nature and the Experience of Modernity." *Body and Society* 6(3–4):58–80.

Leyton, Elliott. 1998. *Touched by Fire: Doctors without Borders in a Third World crisis.* Toronto: McClelland & Stewart.

Lianos, Michalis, with Mary Douglas. 2000. "Dangerization and the End of Deviance: The Institutional Environment." *British Journal of Criminology* 40: 261–78.

Lindholm, Charles. 1990. *Charisma.* Oxford: Blackwell.

Littlewood, Ian. 2001. *Sultry Climates: Travel & Sex.* Cambridge: Perseus.

Livingston, Jonathan. 1998. "Modern Subjectivity and Consumer Culture." In *Getting and Spending: European and American Consumer Societies,* eds. Susan

Strasser, Charles McGovern, and Matthias Judt. Cambridge: Cambridge University Press.

Lledo, Pablo. 2001. "Nuevos productos y servicios reposicionan el sector turístico de Mendoza." *Los Andes.* 18 June 2001. <http://www.losandes.com.ar/asp?nrc=27428>.

London, Charmian. 1915. *The Log of the Snark.* New York: Macmillan.

London, Jack. 1911. *The Cruise of the Snark.* New York: Macmillan.

Lorrain, Claire. 2000. "Cosmic Reproduction, Economics and Politics Among the Kulina of Southwest Amazonia." *Journal of the Royal Anthropological Institute* 6:293–310.

Lubbock, Sir John. 1865. *Prehistoric Times, as Illustrated by Ancient Remains, and the Manners and Customs of Modern Savages.* London: Williams and Norgate.

———— 1870. *The Origin of Civilisation and the Primitive Condition of Man.* London: Longmans, Green & Co.

Luig, Ute. 1996. "Wandarbeiter als Helden." *Historische Anthropologie* 4(3):359–82.

Lukàcs, Georg. 1971. *History and Class Consciousness,* trans. R. Livinstone. Cambridge: MIT Press.

Lupoff, Richard A. 1965. *Edgar Rice Burroughs: Master of Adventure.* New York: Ace Books.

Lush, David. 1993. *Last Steps to Uhuru.* Windhoek: New Namibia Books.

MacCannell, Dean. 1989 [1976]. *The Tourist: A New Theory of the Leisure Class.* New York: Schocken.

———— 1999. *The Tourist: A New Theory of the Leisure Class.* Berkeley: University of California Press.

Mader, Elke. 1999. *Metamorfosis del Poder: Persona, Mito, y Visión en la Sociedad Shuar y Achuar (Ecuador, Perú).* Quito: Ediciones Abya-Yala.

Mader, Elke, and Richard Gippelhauser. 2000. "Power and Kinship in Shuar and Achuar Society," in *Dividends of Kinship. Meaning and Uses of Social Relatedness,* ed. Peter Schweitzer. London: Routledge.

Magnani, Dr. Alfredo Eduardo, and Luis Alberto Parra. 1981. *Aconcagua Argentina.* Mendoza: Ediciones Dhaulagiri.

Maine, Henry Sumner. 1913 [1861]. *Ancient Law: Its Connection with the Early History of Society and Its Relation to Modern Ideas.* London: George Routledge & Sons.

Malinowski, Bronislaw. 1984 [1922]. *Argonauts of the Western Pacific.* Prospect Heights, Ill.: Waveland.

Mandel, Paul. 1963. "Tarzan of the Paperbacks." *Life* 55(22) November 29, no pagination.

Manganaro, Marc. 1992. *Myth, Rhetoric, and the Voice of Authority: A Critique of Frazer, Eliot, Frye and Campbell.* New Haven: Yale University Press.

Marchetti, Gina. 1989. "Action-Adventure as Ideology." *Cultural Politics in Contemporary America,* eds. Ian Angus and Sut Jhally. New York: Routledge.

Maren, Michael. 1997. *The Road to Hell: the Ravaging Effects of Foreign Aid and International Charity.* New York: Free Press.

Marlowe, John. 1967. *Late Victorian: The Life of Sir Arnold Talbot Wilson.* London: The Cresset Press.

Martin, Emily. 1994. *Flexible Bodies: Tracking Immunity in American Culture From the Days of Polio to the Age of AIDS.* Boston: Beacon Press.

Marx, Karl. 1978 [1844]. "Economic and Philosophical Manuscripts of 1844." In *The Marx-Engels Reader,* ed. Richard Tucker. New York: Norton.

Mathers, Kathryn. 2003. *American Travelers and the South African Looking Glass: Learning to Belong in America.* Ph.D. dissertation University of California, Berkeley.

Mayo, William Starbuck. 1849. *Kaloolah; or, Journeyings to the Djebel Kurmi; An Autobiography of Jonathan Romer.* New York: Putnam.

McCracken, Grant. 1990. *Culture and Consumption.* Bloomington: University of Indiana Press.

McLaren, Deborah. 2003. *Rethinking Tourism and Ecotravel.* 2nd Edition. Bloomfield, CT: Kumarian.

McLennan, John F. 1970 [1865]. *Primitive Marriage: An Inquiry into the Origin of the Form of Capture in Marriage Ceremonies.* Chicago: University of Chicago Press.

Mead, Margaret. 1928. *Coming of Age in Samoa: A Psychological Study of Primitive Youth for Western Civilization.* New York: William Morrow & Co.

Meisch, Lynn. 1995. "Gringas and Otavelenos: Changing Tourist Relations." *Annals of Tourism Research* 2292:441–62.

——— 2002. "Sex and Romance on the Trail in the Andes: Guides, Gender and Authority." In *Gender/Tourism/Fun(?),* eds. Margaret Byrne Swain and Janet Henshall. New York: Cognizant Communications Corp.

Messner, Reinhold. 1980. *K2: Mountain of Mountains.* London: Kaye and Ward.

Meyer, Carla. 2002. "Crikey—Irwin's just wild about nature." *San Francisco Chronicle.* July 10th, p. D1.

Miller, Edward. 2000. "Fantasies of Reality: Surviving Reality-Based Programming." *Social Policy* Fall:6–15.

Mills, Sara. 1991. *Discourses of Difference: An Analysis of Women's Travel Writing and Colonialism.* London: Routledge.

Mitchell, Richard G. 1983. *Mountain Experience: the psychology and sociology of Adventure.* Chicago: University of Chicago Press.

——— 2002. *Dancing at Armageddon.* Chicago: University of Chicago Press.

Mitman, Greg. 1999. *Reel Nature: America's Romance with Wildlife on Film.* Cambridge: Harvard University Press.

Moorhead, Alan. 1960. *The White Nile.* London: Fontana.

Morgan, Lewis Henry. 1871. *Systems of Consanguinity and Affinity of the Human Family.* Washington, DC: Smithsonian Institution.

———— 1877. *Ancient Society; or, Researches in the Lines of Human Progress from Savagery through Barbarism to Civilization.* New York: Henry Holt.

Morton, William. 1993. "Tracking the Sign of Tarzan: Trans-Media Representations of a Pop-Culture Icon." In *You Tarzan: Masculinity, Movies and Men,* eds. Pat Kirkham and Janet Thumin. New York: St. Martin's Press.

Munt, Ian. 1994. "Eco-tourism or ego-tourism?" *Race & Class* 36(1):49–60.

Mwarania, Ben. 1999. *Kenya Battalion in Namibia.* Nakuru: Media Document Supplies.

Narayan, Kirin. 1993. "How Native is a 'Native' Anthropologist?" *American Anthropologist* 95(3): 671–86.

Needham, Rodney. 1977. "Rodney Needham." *The Times Literary Supplement* 21, January: 67.

———— 1983. "Tarzan of the Apes: A Re-appreciation." In *Foundation* 28:20–28.

Neirotti, Lisa Delpy. 2003. "An Introduction to Sport and Adventure Tourism." In *Sport and Adventure Tourism,* ed. Simon Hudson. New York: The Haworth Hospitality Press.

Nerlich, Michael. 1987. *Ideology of Adventure: Studies in Modern Consciousness, 1100–1750.* (2 vols.), trans. Ruth Crowley. Minneapolis: University of Minnesota Press.

New York Times. "Excerpts From Pentagon's Plan: 'Prevent Re-Emergence of a New Rival.'" 7 March 1992.

Newman, John, and Michael Unsword. 1984. *Future War Novels: An Annotated Bibliography of Works Published in English since 1946.* Phoenix: Oryx Press.

Nichols, Bruce. 1994. *Blurred Boundaries: Questions of Meaning in Contemporary Culture.* Bloomington: Indiana University Press.

Nye, Russell. 1970. *The Unembarrassed Muse.* New York: Dial Press.

O'Barr, William Mack. 1994. *Culture and the Ad.* Boulder: Westview.

Okely, Judith and Helen Callaway, eds. 1992. *Anthropology and Autobiography.* London: Routledge.

Orlando, Eugene. 2000. "Kanaima." In *Life on Victoria's Street. An Anthology of Victorian Short Stories and Novellas.* [www.eugeneorlando.com/kanaima.htm]

Ortner, Sherry B. 1999. *Life and Death on Mt. Everest: Sherpas and Himalayan Mountaineering.* Princeton: Princeton University Press.

Orvell, Miles. 1989. *The Real Thing.* Chapel Hill: University of North Carolina Press.

Overbey, Mary Margaret. 2001. "AAA Responds to Human Research Oversight Proposals." *Anthropology News* 42(4):10, April.

Paine, Robert. 2002. "Danger and the No-risk thesis." In *Catastrophe and Culture,* eds. Susanna Hoffman and Anthony Oliver-Smith. Sante Fe: School of American Research.

Parfrey, Adam, ed. 2003. *It's a Man's World: Men's Adventure Magazines, the Postwar Pulps.* London: Feral House.

Pasternak, Charles. 2003. *Quest: the Essence of humanity.* London: Wiley.

Peace Corps. 1999. *Peace Corps, the Great Adventure.* Washington, DC: Peace Corps.

Peace Corps. 2001. *Redefine your world.* Washington, DC: Peace Corps.

Peace Corps. 2004. "Why should I volunteer?" Internet document, accessed on 1 April http://www.peacecorps.gov/index.cfm?shell=learn.whyvol,.

Pelton, Robert Young. 1998. *Fielding's The World's Most Dangerous Places.* 3rd Edition Redondo Beach: Fieldings.

Perito, Robert. 2004. *Where Was the Lone Ranger When We Needed Him?* Washington, DC: US Peace Institute.

Perkins, Harvey C. and David C. Thorns. 2001. "Gazing or Performing? Reflections on Urry's tourist gaze in the context of contemporary experience in the antipodes." *International Sociology* 16(2):185–204.

Petit, Pascale. 1998. *Heart of a Deer.* London: Enitharmon Press.

Pettman, Jan Jindy. 1997. "Body Politics: international sex tourism." *Third World Quarterly* 18 (1):93–108.

Phillips, David Clayton. 1996. *Art for Industry's Sake: Halftone Technology, Mass Photography and the Social Transformation of American Print Culture, 1880–1920.* Ph.D. dissertation, Yale University.

Phillips, Richard. 1997. *Mapping Men and Empire: A geography of adventure.* New York: Routledge.

Phillips, Ruth and Christopher Steiner, eds. 1999. *Unpacking Culture: Art and Commodity in colonial and Post Colonial Worlds.* Berkeley: University of California Press.

Piot, Charles. 2002. *Remotely Global.* Chicago: University of Chicago Press.

Pleumarom, Anita. 2001. "Eco-tourism: A new 'Green Revolution' in the third World." Accessed on September 21st at http://www.twnside.org.sg/title/eco2.htm.

Polman, Linda. 2003. *We did nothing : why the truth doesn't always come out when the UN goes in.* London: Viking.

Porges, Irwin. 1975. *Edgar Rice Burroughs: The Man Who Created Tarzan* (2 vols.). New York: Ballantine Books.

Post, Paul. 1996. "The Modern Pilgrim." *Pilgrimage,* eds. Virigilio Elizondo and Sean Freyne. London: SCM Press.

Pratt, Mary Louise. 1992. *Imperial Eyes: Travel Writing and Transculturation.* London; New York: Routledge.

Price, Sally. 1989. *Primitive Art in Civilized Places.* Chicago: University of Chicago Press.

Prins, Harald. 1992. "Bwana Piccer: Martin Johnson (1884–1937) as Ethnographic Film Pioneer." Paper presented to the annual meetings of the American Anthropological Association.

Pritchard, Annette, and Nigel J. Morgan. 2000. "Privileging the Male Gaze: Gendered Tourism Landscapes." *Annals of Tourism Research* 27 (4):884–905.

Radcliffe-Brown, A.R. 1987 [1925]. Appendix C: Letter from A.R. Radcliffe-Brown to Winifred Hoernlé dated 11 August 1925. In *Trails in the Thirstland: The Anthropological Field Diaries of Winifred Hoernlé*, eds. Peter Carstens, Gerald Klinghardt, and Martin West. Cape Town: Centre for African Studies, University of Cape Town.

Raglan, Lord. 1979 (1936). *The Hero*. London: Methuen.

Randis, Alejandro. 1991. *El Aconcagua: El Centinela de Piedra*. Mendoza: Zeta Editores.

—————— 2004. "Récord de visitantes en el Aconcagua." *Los Andes*. 14 Jan. 2004 <http://www.losandes.com.ar/asp?nrc=166793>.

Rappaport, Roy. 1999. *Ritual and Religion in the Making of Humanity*. New York: Cambridge University Press.

Reader, John. 1988. *Missing Links: The Hunt for Earliest Man*. Harmondsworth: Penguin Books.

Reinhard, Johan. 1996. "Peru's Ice Maidens." *National Geographic Magazine* 189(6): 62–81.

Reinhard, Johan. 2001. "A high altitude archaeological survey in northern Chile." *Chungara* 34(1):85–99.

REMHI (Informe Proyecto Interdiocesano de Recuperación de la Memoria Histórica) 1998 *Guatemala Nunca Más* Vol. 3, El Entorno Historico, Guatemala: Oficina de Derechos Humanos del Arzobispado de Guatemala. Available from http://www.odhag.org.gt; INTERNET.

Rice, Laura. 1991. "Nomad Thought: Isabelle Eberhardt and the Colonial Project." *Cultural Critique* 17:151–76.

Riding, Alan. 2004. "Globe-Trotting Englishwomen Who Helped Map the World." <http://www.nytimes.com/2004/08/19/arts/design/19wom...>

Riffenbaugh, Beau. 1992. *The Myth of the Explorer*. New York: Belhaven Press.

Ritchie, Bruce. 2002. "Crocodile Hunter inspires ecologist's own snake documentary." *Milwaukee Journal Sentinel*, 12 July, p. 06B.

Robertson, A.F. 1984. *People and the State: An Anthropology of Planned Development*. Cambridge: Cambridge University Press University of California Press.

Roe, Emery. 1991. "Development Narratives, or Making the Best of Blueprint Development." *World Development* 19 (4):287–300.

—————— 1995. "Except Africa: Postscript to a Special Section on Development Narratives." *World Development* (23)6:1065–69.

Rogin, Michael. 1993. "'Make My Day!': Spectacle as Amnesia in Imperial Politics [and] The Sequel." In *Cultures of United States Imperialism*, eds. Amy Kaplan and Donald Pease. Durham: Duke University Press.

Rojek, Chris. 1994. *Ways of Escape*. Lanham, MD: Rowman & Littlefield.

—————— 2000. *Leisure and Culture*. New York: St Martins Press.

Rojek, Chris, and John Urry. 1997. "Transformations of Travel and Theory." In *Touring Cultures: Transformation of Travel and Theory*, eds. Chris Rojek and John Urry. London: Routledge.

Rosaldo, Renato. 1989. *Culture and Truth*. Boston: Beacon Press.

Roth, Kenneth. 2004. "The Law of War in the War on Terror." *Foreign Affairs*. January, February 83(1):1–10.

Rubenstein, Steven. 2001. "Colonialism, the Shuar Federation, and the Ecuadorian State." *Environment and Planning D: Society and Space* 19(3):263–93.

——— 2002. *Alejandro Tsakimp: A Shuar Healer in the Margins of History*. Lincoln: University of Nebraska Press.

——— 2004. "Fieldwork and the Erotic Economy on the Colonial Frontier." *Signs: Journal of Women in Culture and Society* 29(4)1041–71.

Rumsfeld, Donald. 2002. "Transforming the Military." *Foreign Affairs*. May/June 81(3):20–32.

Ryan, Chris, and C. Michael Hall. 2001. *Sex Tourism: Marginal People and Liminalities*. London: Routledge.

Ryan, Chris, and Rachel Kinder. 1996. "Sex, tourism and sex tourism: fulfilling similar needs?" *Tourism Management* 17(7):507–18.

Sahlins, Marshall. 1976. *Culture and Practical Reason*. Chicago: University of Chicago Press.

Said, Edward W. 1979. *Orientalism*. New York: Vintage Books.

Sala, Germán. 2002. "Nueva apuesta al turismo chileno." *Los Andes* 1 Dec. 2002. http://www.losandes.com.ar/asp?nrc=101390>.

Santos Granero, Fernando. 1986. "Power, Ideology, and the Ritual of Production in Lowland South America." *Man* (N.S.) 21(4): 657–79.

Scheibe, Karl E. 1986. "Self-narratives and Adventure." *Narrative Psychology*. Ed. T.R. Sarbin. Praeger: New York.

Scheinman, Diane. 2000. "Martin Johnson's Consuming Passions: Cannibalism and the Cinema." Paper presented to Visible Evidence VIII Conference, Utrecht, Netherlands.

Schutz, Alfred. 1973. *The Structures of the Life-World*. London: Heinemann.

Schwimmer, Brian and D. Michael Warren, eds. 1993. *Anthropology and the Peace Corps: Case Studies in Career Preparation*. Ames, IA: Iowa State University Press.

Scott, James. 1998. *Seeing Like a State: How Certain Schemes to Improve the Human Condition have Failed*. New Haven: Yale University Press.

Sellerburg, Ann-Mari. 1994. *A Blend of Contradictions: Georg Simmel in Theory and Practice*. New Brunswick and London: Transaction Publishers.

Server, Lee. 2002. *Encyclopedia of Pulp Fiction Writers*. New York: Facts on File.

Seymour-Smith, Charlotte. 1988. *Shiwiar: Identidad Etnica y Cambio en el Rio Corrientes*. Quito-Lima: Abya Yala-CAAP.

——— 1991. "Women Have No Affines and Men No Kin: The Politics of Jivaroan Gender Relations." *Man* (N.S.) 26:629–49.

Shelley, Mary. 1985 [1818]. *Frankenstein; or, The New Prometheus*. London: Penguin.

Shnayerson, Michael. 2002. "The Man Who Walked Through Time." *National Geographic Adventure*, January/February: 70–75, 105–6.

Shriver, Sargent. 1964. *The Point of the Lance.* New York: Harper and Row.

Sillitoe, Alan. 1995. *Leading the Blind: A Century of Guidebook Travel, 1815–1911.* London: Macmillan.

Simmel, Georg. 1950a. "Sociability." In *The Sociology of Georg Simmel,* ed. Kurt Wolff. Glencoe, Ill.: Free Press.

———— 1950b. "The Stranger." In *The Sociology of Georg Simmel,* ed. Kurt Wolff. Glencoe, Ill.: Free Press.

———— 1955 [1908]. *Conflict and the Web of Group Affiliations.* New York: Free Press.

———— 1959. *Georg Simmel, 1858–1918,* ed. Kurt Wolff. Columbus: Ohio State University Press.

———— 1965. "The Adventure." In *Essays on Sociology, Philosophy and Aesthetics by Georg Simmel,* ed. K. Wolff. New York: Harper and Row.

———— 1971. "The Adventure." In *Simmel on Individuality and Social Forms,* ed. D. Levine. Chicago: University of Chicago Press.

———— 1983/1997. "The Adventure." In *Simmel on Culture,* eds. David Frisby and Mike Featherstone. London: Sage.

———— 1984. *On Women, Sexuality and Love,* trans. Guy Oakes. New Haven: Yale University Press.

Simpson, Sarah. 2001. "Full of Croc?" *Scientific American* 284(4). WWW site accessed 11 July 2001 http://sciam.com/2001/0401issue0401scicit3box1.html.

Smith, Barbara. 2001. "Rescuing the Ciphers." *Christian Science Monitor* 9 March: 11.

Smith, Michael with Moreen Dee. 2003. *Peacekeeping in East Timor: The Path to Independence.* Boulder: Lynne Reiner.

Smith, Sidonie. 1995. "Isabelle Eberhardt Travelling Other/wise: The 'European' Subject in 'Oriental' Identity." In *Encountering the Others.* ed. G. Brinker-Gabler. Albany: State University of New York Press.

Snyder, Gary. 1990. *The Practice of the Wild.* San Francisco: North Point Press.

South African Defence Force. 1977. *Ethnology Manual for the Soldier.* Mimeograph.

Spurr, David. 1993. *The Rhetoric of Empire: Colonial Discourse in Journalism, Travel Writing and Imperial Administration.* Durham & London: Duke University Press.

Stagl, Justin. 1990. "The Methodology of Travel in the 16th Century." *History and Anthropology* 4:303–338.

Stainton, John, director. 2000. *Crocodiles of the Revolution.* Discovery Channel/Animal Planet.

———— director. 2002. *Crocodile Hunter: Collision Course.* MGM/UA Studios.

Stark, Peter. 2001. *Last Breath: the Limits of Adventure.* New York: Ballantine Books.

Starn, Orin. 1999. *Nightwatch: The Politics of Protest in the Andes.* Durham, North Carolina: Duke University Press.

Steel, Daniel. 1999. "Trade Goods and Jívaro Warfare: The Shuar 1850–1957, and the Achuar, 1940–1978." *Ethnohistory* 46(4): 745–76.

Stirling, M.W. 1938. *Historical and Ethnological Material on the Jívaro Indians.* Washington, D.C.: United States Printing Office.

Stocking, George W. 1971. "What's in a Name? The Origins of the Royal Anthropological Institute: 1837–1871." *Man* (n.s.), 6: 369–90.

———— 1989. "The Ethnographic Sensibility of the 1920s and the Dualism of the Anthropological Tradition." *Romantic Motives: Essays in Anthropological Sensibility,* ed. George W. Stocking. Madison: University of Wisconsin Press.

Stoll, David. 1993. *Between Two Armies in the Ixil Towns of Guatemala.* New York: Columbia University Press.

Stoller, Paul. 1997. *Sensuous Scholarship.* Philadelphia: University of Pennsylvania Press.

Strachan, Alexander. 2002. "Die 'Werklikheid' van die Grens." In *De Helende Kracht van Literatuur,* eds. Chris van der Merwe and Rolf Wolfswinkel. Haarlem: In de Knipscheer.

Strain, Ellen. 2003. *Public Places, Private Spaces: Ethnography, Entertainment and the Tourist Gaze.* New Brunswick: Rutgers University Press.

Street, Brian V. 1975. *The Savage in Literature: Representations of "Primitive" Society in English Fiction, 1858–1920.* London: Routledge & Kegan Paul.

Sumich, Jason. 2002. "Looking for the 'other': tourism, power, and identity in Zanzibar." *Anthropology Southern Africa.* 25(1 & 2). 39–45.

Suttles, Gerald. 1983. "Foreword." In *Mountain Experience: the Psychology and Sociology of Adventure.* Richard G. Mitchell. Chicago and London: University of Chicago Press.

Swarbrooke, John, et al. eds.. 2003. *Adventure Tourism: The New Frontier.* Oxford: Butterworth-Heinemann.

Taplin, Thomas E. 1992. *Aconcagua: The Stone Sentinel, Perspectives of an Expedition.* Santa Monica: Eli Ely Publishers.

———— 2000. *TurPlan 2000–2005, Plan Estratégico de Desarrollo Turístico de Mendoza.* Subsecretaría de Turismo de la Provincia.

Taylor, Anne-Christine. 1981. "God-Wealth: The Achuar and the Missions." In *Cultural Transformations and Ethnicity in Modern Ecuador,* ed. Norman E. Whitten, Jr. Urbana: University of Illinois Press.

———— 1993. "Remembering to Forget: Identity, Mourning and Memory Among the Jívaro." *Man* 28:653–78.

———— 2001. "Wives, Pets, and Affines: Marriage Among the Jívaro." In *Beyond the Visible and the Material: The Amerindianization of Society in the Work of Peter Rivière,* eds. Laura Rival and Neil Whitehead. Oxford: Oxford University Press.

Taylor, Jacqueline Sanchez. 2001. "Dollars are a Girl's Best Friend? Female Tourists' Sexual Behaviour in the Caribbean." *Sociology* 35 (3):749–64.

Terry, Fiona. 2002. *Condemned to Repeat? The Paradox of Humanitarian Action*. Ithaca: Cornell University Press.

Theroux, Paul. 1998 [1967]. "Tarzan was an expatriate." *Transition* 75/76:46–58.

Thomas, Lowell. 1976. *Good Evening Everybody*. New York: William Morrow.

Thomas, Nicholas. 1987. "Narrative as Practice?" *Anthropology Today* 3(5):8–11.

Thornberry, Cedric. 1990. *The Untag Experience in Namibia—First Phase*. Johannesburg: South African Institute of International Affairs.

Tiger, Lionel. 1969. *Men in Groups*. New York: Random House.

Todorov, Alexander and Anesm Mandisodza. 2003. "Public Opinion on Foreign Policy: The Multilateral Public that Perceives Itself as Unilateral." *Policy Brief*, Washington DC: Woodrow Wilson School of Public and International Affairs.

Torgovnick, Mariana. 1990. *Gone Primitive*. Chicago: University of Chicago Press.

Travel Industry Association of America. 1998. *Adventure Travel Report 1997*. Accessed at http://www.tia.org/Pubs/toc_advent97.asp

Trilling, Lionel. 1972. *Sincerity and Authenticity*. Cambridge: Harvard University Press.

Turner, Frederick Jackson. 1920 [1893]. "The Significance of the Frontier in American History." In *The Frontier in American History*. New York: Henry Holt and Company.

Turner, Terence. 1995. "Social Body and Embodied Subject: Bodiliness, Subjectivity, and Sociality among the Kayapo." *Cultural Anthropology* 10(2):143–170.

Turner, Victor W. 1967. *The Forest of Symbols: aspects of Ndembu ritual*. Ithaca: Cornell University Press.

——— 1969. *The Ritual Process: structure and anti-structure*. Chicago: Aldine.

——— 1972. *Dramas, Fields, and Metaphors: symbolic action in human society*. Ithaca: Cornell University Press.

——— 1979. *Process, Performance, and Pilgrimage: a study in comparative symbology*. New Delhi: Concept.

Tylor, Edward Burnett. 1871. *Primitive Culture: Researches into the Development of Mythology, Philosophy, Religion, Art and Custom* (2 vols.). London: H. Murray.

United Nations. 1990. *The Blue Helmits*. New York: United Nations.

UNTAG. 1990. *Untag in Namibia: A New Nation is Born*. Windhoek: UNTAG.

Up de Graff, Fritz W. 1923. *Head Hunters of the Amazon: Seven Years of Exploration and Adventure*. Garden City: Garden City Publishing Company, Inc.

Urry, John. 1990. *The Tourist Gaze: Leisure and Travel in Contemporary Societies*. London: Sage Publications.

Van den Berghe, Pierre. 2002. "Risk and deceit in transient, non-repeated interactions: the case of tourism." In *Risky Transactions*, ed. Frank Salter. New York: Berghahn.

van Vuuren, Chris. 2000. "Why We Climb, Trek and Backpack: Notions of Pilgrimage and Ritual and the Quest for the Other." *Journal of the Mountain Club of South Africa* 103:8–13.

Venter, Al J. 1983. "Chipping Away at SWAPO." *Soldier of Fortune Magazine* January:46–51.

Verne, Jules. 1871 [French edition, 1864]. *A Journey to the Centre of the Earth*. London: Griffith & Farran.

Verwoerd, Wilhelm. 2002. "Oorlog, trauma en die WVK." In *De Helende Kracht van Literatur* eds. Chris van der Merwe and Rolf Wolfswinkel. Haarlem: In de Knipscheer.

Vidal, Gore. 1963. "Tarzan Revisited." *Esquire*. December. 192–93.

Vines, Stuart. 1950. "The Fitz Gerald Expedition (1897): Tupungato Ascended." In *Challenge: An Anthology of the Literature of Mountaineering,* ed. William Robert Irwin. New York: Columbia University Press.

Vivanco, Luis. 2001. "Totally Lost: Entertaining Ourselves with the Ugly American." *American Anthropology Newsletter,* November 2001.

——— 2002. "Seeing Green: Knowing and Saving the Environment on Film." *American Anthropologist* 104(4):1195–1204.

——— 2004. "The Work of Environmentalism in an Age of Televisual Adventures." *Cultural Dynamics* 16 (1):93–115.

Walker, Marina. 2003a. "Las mujeres pisan fuerte en el Aconcagua." *Los Andes* 3 Feb. 2003. <http://www.losandes.com.ar/asp?nrc=

——— 2003b. "Termina una temporada con récord de turistas en el Aconcagua" 2003. *Los Andes* 13 Mar. 2003. < http://www.losandes.com.ar/asp?nrc=115434.

Wanderer, Jules J. 1987. "Simmel's Forms of Experiencing: The Adventure as Symbolic Work." *Symbolic Interaction* 10 (1):21–28.

Wardle, Huon. 2002. "Jamaican Adventures: Simmel, Subjectivity and Extraterritoriality in the Caribbean." *Journal of the Royal Anthropological Institute* (NS) 5:523–39.

Weber, Max. 1958. *The Protestant Ethic and the Spirit of Capitalism*. New York: Scribners.

Weisinger, Herbert. 1961."The Branch That Grew Full Straight." *Daedalus* Volume 90/2: 388–99.

White House, 2002. "The National Security Strategy of the United States of America." September. http://usinfo.state.gov/topical/poll/terror/secstrat.htm

White, Randy Wayne. 1997. "Croco%#@! Dundee." *Outside Magazine* November. Accessed on 20 March 2001. http://www.outsidemagazine.com/magazine/1197/9711out.html.

Whitehead, Neil L. 2002. *Dark Shamans. Kanaimà, and the Poetics of Violent Death*. Durham: Duke University Press.

Wilson, Sir Arnold T. 1940. *S.W. Persia: Letters and Diary of a Young Political Officer*. Oxford: Oxford University Press.

Withey, Lynne. 1997. *Grand Tours and Cook's Tours*. New York: Morrow.

Wright, David, producer/cinematographer. 2002. *Quest for the Rainbow Serpent.* National Geographic Explorer Films.

Yengoyan, Aram A. 2002. "Simmel, Modernity, and Germanisms." *Comparative Studies in Society and History* 44:620–25.

—— 2003. "Lyotard and Wittgenstein and the Question of Translation." In *Translating Cultures: Perspectives on Translation and Anthropology,* eds. Paula G. Rubel and Abraham Rosman. Oxford: Berg.

—— 2004. "Whatever Happened to the Soul?" *Comparative Studies in Society and History* 46:411–17.

Zabaleta, Martin. 2004. Personal comment, at interviewee's home. Bozeman, MT, USA.

Zajonc, Robert B. 1952. "Aggressive Attitudes of the 'Stranger' as a Function of Conformity Pressure." *Human Relations* 5 (May): 205–16.

Zorrilla, Juan José. 1997. "El centenario del centinela de piedra." *Desnivel, Revista de Montaña,* No. 126: 28–43.

Zuckerman, Marvin. 1994. *Behavioral Expressions and Biosocial Bases of Sensation Seeking.* New York: Cambridge University Press.

Zur, Judith. 1998. *Violent Memories: Mayan War Widows in Guatemala.* Boulder, Colorado: Westview Press.

Zurick, David. 1995. *Errant Journeys: Adventure Travel in a Modern Age.* Austin: University of Texas Press.

Zweig, Paul. 1974. *The Adventurer.* Princeton: Princeton University Press.

Index

Related Titles of Interest

GREEN ENCOUNTERS
Shaping and Contesting Environmentalism in Rural Costa Rica
Luis A. Vivanco

Since the 1970s and 1980s, Monte Verde, Costa Rica has emerged as one of the most renowned sites of nature conservation and ecotourism in Costa Rica, and some would argue, Latin America. It has received substantial attention in literature and media on tropical conservation, sustainable development, and tourism. Yet most of that analysis has uncritically evaluated the Monte Verde phenomenon, using celebratory language and barely scratching the surface of the many-faceted socio-cultural transformations provoked by and accompanying environmentalism. Because of its stature, Monte Verde represents an ideal case study to examine the socio-cultural and political complexities and dilemmas of practicing environmentalism in rural Costa Rica.

Based on many years of close observation, this book offers rich and original material on the ongoing struggles between environmental activists and of collective and oppositional politics to Monte Verde's new "culture of nature."

Luis A. Vivanco is Assistant Professor of Anthropology at the University of Vermont.

224 pages, 7 ills, 3 maps, bibliog., index
ISBN 1-84545-168-6 Hb $75.00/£45.00
Volume 3, Environmental Anthropology and Ethnobiology

TOURISM
Between Place and Performance
Edited by Simon Coleman and Mike Crang

Many accounts of tourism have adopted an almost paradigmatic visual model of the gaze. This collection presents an expanded notion of spectatorship with a more dynamic sense of embodied and performed engagement with places. The approach resonates with ideas in anthropology, sociology, and geography on performance, invented traditions, constructed places and traveling cultures. Contributions highlight the often contradictory, contested and paradoxical constructions of landscape and community involved both in tourist attractions and among tourists themselves. The collection examines many different practices, ranging from the energetic pursuit of adventure holidays to the reading of holiday brochures. It illustrates different techniques of seeing the landscape and a variety of ways of creating and performing the local. Chapters thus demonstrate the mutual entanglement of practices, images, conventions, and creativity. They chart these global flows of people, texts, images, and artefacts. Case studies are drawn from diverse types of tourism and destination focused around North America, Europe, and Australasia.

Simon Coleman teaches in the Department of Anthropology, University of Durham. Mike Crang is Lecturer in the Department of Geography, University of Durham.

2002. 240 pages, bibliog., index
ISBN 1-57181-746-8 Pb $25.00/£15.00
ISBN 1-57181-745-X Hb $59.95/£40.00

Berghahn Books, Inc. 150 Broadway, Suite 812, New York, NY 10038, USA

Berghahn Books, Ltd. 3 Newtec Place, Magdalen Rd. Oxford OX4 1RE, UK

orders@berghahnbooks.com   www.berghahnbooks.com

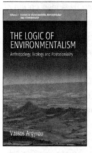

Lightning Source UK Ltd.
Milton Keynes UK
UKOW051037201212

203939UK00008B/192/P